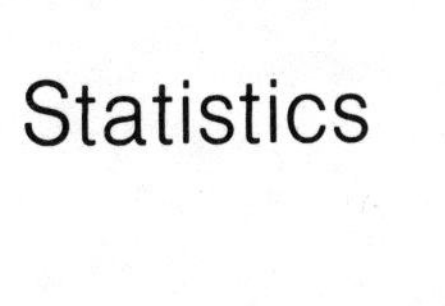
Statistics

## The Manchester Physics Series

*General Editors*

**F. MANDL R. J. ELLISON D. J. SANDIFORD**

*Physics Department, Faculty of Science, University of Manchester*

| | |
|---|---|
| **Properties of Matter:** | B. H. Flowers and E. Mendoza |
| **Optics:** *Second Edition* | F. G. Smith and J. H. Thomson |
| **Statistical Physics:** *Second Edition* | F. Mandl |
| **Solid State Physics:** | H. E. Hall |
| **Electromagnetism:** | I. S. Grant and W. R. Phillips |
| **Atomic Physics:** | J. C. Willmott |
| **Electronics:** | J. M. Calvert and M. A. H. McCausland |
| **Electromagnetic Radiation:** | F. H. Read |
| **Statistics:** | R. J. Barlow |

*In preparation*

| | |
|---|---|
| **Particle Physics:** | B. R. Martin and G. Shaw |

# STATISTICS

*A Guide to the Use of Statistical Methods in the Physical Sciences*

**Roger Barlow**
*Department of Physics*
*Manchester University*



John Wiley & Sons

CHICHESTER NEW YORK BRISBANE TORONTO SINGAPORE

*Other Wiley Editorial Offices*

John Wiley & Sons, Inc., 605 Third Avenue,
New York, NY 10158-0012, USA

Jacaranda Wiley Ltd, G.P.O. Box 859, Brisbane,
Queensland 4001, Australia

John Wiley & Sons (Canada) Ltd, 22 Worcester Road,
Rexdale, Ontario M9W 1L1, Canada

John Wiley & Sons (SEA) Pte Ltd, 37 Jalan Pemimpin 05-04,
Block B, Union Industrial Building, Singapore 2057



***Library of Congress Cataloging-in-Publication Data***

Barlow, Roger (Roger J.)
Statistics: a guide to the use of statistical methods in the physical sciences.

(The Manchester physics series)
Bibliography: p.
Includes index.
1. Statistical physics. I. Title. II. Series.
QC174.8.B365 1989 530.1′595 88-33908
ISBN 0 471 92294 3
ISBN 0 471 92295 1 (pbk.)

***British Library Cataloguing in Publication Data***

Barlow, Roger
Statistics: a guide to the use of statistical methods in the physical sciences.
1. Statistical mathematics
I. Title II. Series
519.5

ISBN 0 471 92294 3
ISBN 0 471 92295 1 pbk

Phototypesetting by Thomson Press (India) Limited, New Delhi, India.
Printed and bound in Great Britain by Courier International Ltd, Tiptree, Essex

*To my father*

# Editors' Preface to the Manchester Physics Series

The first book in the Manchester Physics Series was published in 1970, and other titles have been added since, with total sales world-wide of more than a quarter of a million copies in English language editions and in translation. We have been extremely encouraged by the response of readers, both colleagues and students. The books have been reprinted many times, and some of our titles have been rewritten as new editions in order to take into account feedback received from readers and to reflect the changing style and needs of undergraduate courses.

The Manchester Physics Series is a series of textbooks at undergraduate level. It grew out of our experience at Manchester University Physics Department, widely shared elsewhere, that many textbooks contain much more material than can be accommodated in a typical undergraduate course and that this material is only rarely so arranged as to allow the definition of a shorter self-contained course. In planning these books, we have had two objects. One was to produce short books: so that lecturers should find them attractive for undergraduate courses; so that students should not be frightened off by their encyclopaedic size or their price. To achieve this, we have been very selective in the choice of topics, with the emphasis on the basic physics together with some instructive, stimulating and useful applications. Our second aim was to produce books which allow courses of different length and difficulty to be selected, with emphasis on different applications. To achieve such flexibility we have encouraged authors to use flow diagrams showing the logical connections between different chapters and to put some topics in starred sections. These cover more advanced and alternative material which is not required for the understanding of later parts of each volume. Although these books were conceived as a series, each of them is self-contained and can be used independently of the others. Several of them are suitable for wider use in other sciences. Each author's preface gives details about the level, prerequisites, etc., of his volume.

We are extremely grateful to the many students and colleagues, at Manchester and elsewhere, whose helpful criticisms and stimulating comments have led to many improvements. Our particular thanks go to the authors for all the work they have done, for the many new ideas they have contributed, and for discussing patiently, and often accepting, our many suggestions and requests. We would also like to thank the publishers, John Wiley & Sons, who have been most helpful.

F. Mandl
R. J. Ellison
D. J. Sandiford

*January, 1987*

*The generall end therefore of all the book is to fashion*
*a noble person in vertuous and gentle discipline*

—*Edmund Spencer*

# Author's Preface

Many science students acquire a distinctly negative attitude towards the subject of statistics. The reasons for this are clear. The traditional first year concentrated statistics course of derivations and exhortations makes little impact on the young undergraduates, who want to get to grips with the basic truths of their chosen subject and have no interest in sordid details like error bars. The hapless students then go to laboratory classes, in which their enjoyment of the experiments is marred by the awful chore of the 'error analysis' at the end, where, whatever they do, they inevitably get harshly criticised for doing it wrong. Under such circumstances, 'statistics' can soon become a collection of meaningless ritual, to be gone through correctly if harsh words and bad marks are to be avoided.

As a student I was no different from any other in this respect. But later, in the real world, doing real experiments, statistics began to matter. Over the years I got to grips with the subject, by talking to colleagues and digging in reference books, and was agreeably surprised to discover that it had an internal logic and structure. Once one really got into it, it made sense. Eventually the time came when people started asking me questions, and I somehow acquired a reputation as the local statistics expert. On this basis I devised a course, which was given as a set of lectures to students at

Manchester University. This has convinced me that statistics can be taught to students in such a way as to make it interesting for them, and give them a real grasp of the subject.

This book has grown out of the lecture notes given out with the course. Despite the shelves full of books on 'statistics' in any library or university bookshop, there is a desperate lack of any suitable textbook for the physical sciences beyond the very elementary level. The books available are mainly aimed at the biological and social sciences; for those of us in other fields they are inappropriate, both in content and treatment. They deal largely with samples and surveys, and the problems of hypothesis testing, whereas we are more concerned with the theory of measurements and errors, and with the problem of estimation. Furthermore they assume, usually correctly, that those for whom they are intended (geographers, psychologists, and suchlike) will fear and loathe anything at all mathematical. They therefore avoid anything beyond (or even, in some cases, including) the most elementary algebra. Now, although physicists and chemists may fight shy of high-powered abstract mathematics, they can happily differentiate and integrate simple functions and follow basic algebra. They are thus entitled to a reasonable explanation of the mathematics involved in statistical calculations, and able to benefit from it. This book thus assumes a reasonable degree of numeracy from the reader, but nothing outstanding—any real mathematician will find it hopelessly naive and unrigorous.

This book is thus the textbook I would like to have had available, both as a student and when teaching students, and for my own use with real problems. I hope that others will find it useful and interesting, and that it will eventually lead them not only to use and understand statistics, but to enjoy it.

I would like to record my acknowledgements to the many people who, by discussions and advice, have helped form my ideas on the subject, to the students on my course for acting as guinea-pigs for the material, to John Ellison for many helpful comments in preparing the manuscript for publication, and finally to my wife Ann for putting up with the trials of a traumatic author with patience and understanding.

ROGER BARLOW
*Manchester*

*4 October 1988*

# Contents

*'It's not the figures themselves,' she said finally, 'it's what you do with them that matters'*

*—K.A.C. Manderville*

CHAPTER 1

# Using Statistics

Statistics is a tool. In experimental science you plan and carry out experiments, and then analyse and interpret the results. To do this you use statistical arguments and calculations. Like any other tool—an oscilloscope, for example, or a spectrometer, or even a humble spanner—you can use it delicately or clumsily, skilfully or ineptly. The more you know about it and understand how it works, the better you will be able to use it and the more useful it will be.

The fundamental laws of classical science do not deal with statistics or errors. Newton's law of gravitation, for example, reads

$$F = \frac{GMm}{r^2}$$

in pure and beautiful simplicity. The figure in the denominator is given as 2—exactly 2, not $2.000 \pm 0.012$ or anything messy like that. This can lead people to the idea that statistics has nothing to do with 'real' scientific knowledge.

But where do the laws come from? Newton's justification came from the many detailed and accurate astronomical observations of Tycho Brahe and others. Likewise Ohm's law

$$V = IR$$

which appears so straightforward and elementary to us today, was based on Ohm's many careful measurements with primitive apparatus. When you are *studying* science you may find no use for statistics—until you meet quantum mechanics, but that is another story—but as soon as you begin *doing* science, and want to know what measurements really mean, it becomes a matter of vital importance.

This is a textbook on statistics for the physical sciences. It treats the subject from the basic level up to a point where it can be usefully applied in analysing real experiments. It aims to cover most situations that are likely to be met with, and also provide a grasp of statistical ideas, terminology, and language, so that more advanced works can be consulted and understood should the need arise. It is thus intended to be usable both as a textbook for students taking a course in the subject, and also as a handbook and reference manual for research workers and others when they need statistical tools to extract their experimental results.

These two modes of use give rise to requirements in the ordering of the material which are not always happily reconcilable. For reference use one wants to group all material on a given topic together, but for teaching purposes this would be like learning a language from a dictionary. The solution adopted is that the unstarred sections cover the material roughly appropriate to a first year undergraduate course. They can sensibly be taken in order, with no anticipation of later material. The starred sections fill in the gaps; they may require knowledge of material in later sections, but when this occurs it is explicitly pointed out. Most of the basic material is in the early chapters, and Chapters 7, 9, and 10 contain entirely higher-level material. First-time-through readers should not be scared or put off by any apparent mathematical complexity they observe in some of the starred sections: these can (and should) be skipped over with a clean conscience, as they are not needed for later unstarred sections of the course.

*'Data! Data! Data!', he cried impatiently.*
*'I can't make bricks without clay'.*

*—Sir Arthur Conan Doyle*

CHAPTER 

# Describing the Data

It all starts with the data. You may call them a *set of results*, or a *sample* or the *events*, but whatever the name, they consist of a set of basic measurements from which you're trying to extract some meaningful information.

To make your data mean something, particularly to an outside audience, you need to display them pictorially, or to extract one or two important numbers. There are many such numbers and ways of presenting the data in graphic form, and this chapter is devoted to methods of describing the data in a useful and meaningful way, without attempting any deeper analysis or inference. This is known as *descriptive statistics.*

## 2.1 TYPES OF DATA

Data[†] are called *quantitative* or *numeric* if they can be written down as numbers, and *qualitative* or *non-numeric* if they cannot. Qualitative data are

[†]Note for pedants: 'data' is a plural noun. Thus one should say 'The data fit...', 'Data were observed...' rather than 'The data fits...', 'Data was observed...'. The singular, never used, is 'datum'.

rather hard to work with as they do not offer much scope for mathematical treatment, so most of the subject of statistics, and likewise most of this book, deals with quantitative, numerical measurements.

Quantitative measurements divide further into two types. Some, by their very nature, have to be integers and these are called *discrete* data. Others are not constrained in this way and their values are real numbers. These are called *continuous* data. Continuous data cannot be recorded exactly, as you cannot write down an infinite number of decimal places. Some sort of *rounding* and loss of precision has to occur.

For example, if you were to examine a sample of motor cars and record their colours, these would be qualitative data. The number of seats in each car has to be an integer, and would be discrete numeric data, as would the number of wheels. The lengths and the weights of the cars would be continuous numeric data.

Usually one of the first things to do in making sense of the data (which is just a pile of raw results) is to divide them into *bins* (also called *groups* or *classes* or *blocks*). For example, the results of tossing 20 coins, each of which comes down either heads (H) or tails (T)

$$\{H, T, H, H, T, H, T, H, H, H, T, T, H, T, T, H, T, H, H, T\}$$

can be written as $\{11H, 9T\}$. This conveys the same information much more clearly and concisely.

For continuous numeric data it is not quite so simple, as your values are (almost certainly) all different, if you use enough decimal places. You have to group together adjacent numbers, using a range of values to define each bin. This means further rounding of values and throwing away precision information, which is the price you pay for rendering the data comprehensible. Usually the bins are chosen to be all the same uniform size, but in some cases it makes sense to use non-uniform bins of different sizes.

For discrete numeric data this grouping together of adjacent values is not compulsory, but it may be desirable when the numbers of data points with any particular value are small.

## 2.2 BAR CHARTS AND HISTOGRAMS

The numbers of events in the bins can be used to draw bar charts (see Figure 2.1) and histograms.

There is a technical difference between a bar chart and a histogram in that the number represented is proportional to the *length* of bar in the former and the *area* in the latter. This matters if non-uniform binning is used. Bar charts can be used for qualitative or quantitative data, whereas histograms can only be used for quantitative data, as no meaning can be attached to the width of the bins if the data are qualitative.

For quantitative, numeric, data, you have to choose the width of the bins

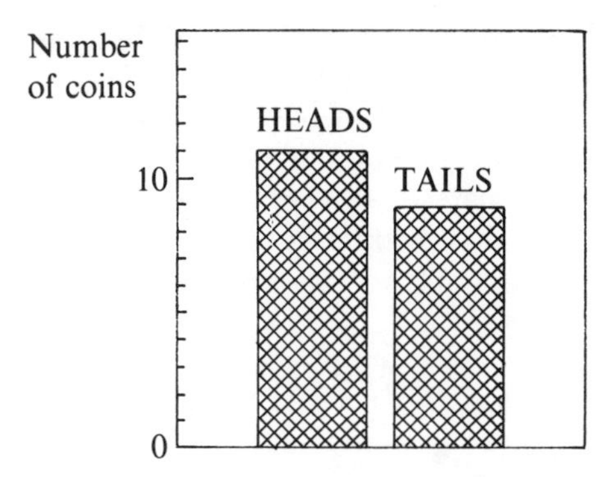


Fig. 2.1. A bar chart displaying the data in the previous section.

to be used in the display (see Figures 2.2). This requires thought. If the bins chosen are too narrow, then there are very few events in each bin, and the numbers are dominated by fluctuations. Ideally there should be at least ten events in each bin, and the more the better. If they are too wide, then real detail can be obscured if the bin stretches over genuine variations in the distribution. Ideally, the difference between contents of adjacent bins should be small. The choice is yours—it is a matter of personal judgement. It may well be, particularly if the number of events is small, that there is no way of satisfying both ideal requirements. In this case you just have to do the best you can with the data available.

There are other ways of representing the data using pictures: ideographs, frequency polygons, pie charts, prismograms, scatter plots, and many more. However, it is not necessary to give you all the details. They are designed to be straightforward to understand, and are therefore straightforward to use. Some people become very excited about 'right' and 'wrong' ways of doing things, and come almost to blows over whether gaps between bars in a bar

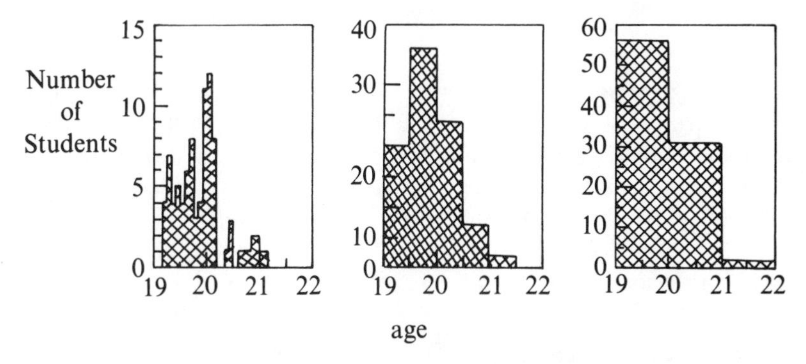


Fig. 2.2 The ages (in years) of a group of second year students, showing the effects of choosing different bin sizes for the same data.

chart are compulsory or optional, and similar trivial matters. Such details are not really important. It is very much a matter of your own taste as to how you display your data, and the scales and axes you use. The object of any description is to convey an idea of your data to your audience in a way that is effective, easy to grasp, and honest. That is all that really matters. (For examples of dishonest methods, consult *How to Lie with Statistics* by Darrell Huff (see Bibliography) or any daily newspaper.)

## 2.3 AVERAGES

### 2.3.1 The Arithmetic Mean

If you want to describe your data with just one number, the best and most meaningful one to use is almost certainly the arithmetic mean. This is denoted by an upper bar over the quantity concerned: thus if there are $N$ elements in the set of data

$$\{x_1, x_2, x_3, \ldots, x_N\}$$

then the mean value of $x$ is

$$\bar{x} = \frac{1}{N}\sum_{i=1}^{N} x_i. \tag{2.1}$$

In the same way you can calculate the mean value of any function $f(x)$:

$$\bar{f} = \frac{1}{N}\sum_{i} f(x_i). \tag{2.2}$$

If the data have been binned, and bin $j$ corresponds to a value $x_j$ and contains $n_j$ data elements, then these means can be written

$$\bar{x} = \frac{1}{N}\sum_{j} n_j x_j \tag{2.3}$$

$$\bar{f} = \frac{1}{N}\sum_{j} n_j f(x_j). \tag{2.4}$$

Note carefully the apparent difference between these formulae and the ones above. Which you use depends on whether you are summing over *elements* of the data set or over *bins*. If rounding has occurred then the average from the bins is less accurate than the average over the unrounded elements, so it is better to use the unrounded data if you can.

*Example Weights of foils*
The weights of 5 metal foils are 25g, 24g, 27g, 29g, and 25g.

The mean weight is therefore $\dfrac{25+24+27+29+25}{5} = 26\text{g}.$

*Example Occupancy of cars*
In a survey of 100 cars passing a checkpoint, 72 contained only 1 occupant (the driver), 23 had 2, 2 had 3, and 3 had 4.

$$\text{Mean number of occupants per car} = \frac{72 \times 1 + 23 \times 2 + 2 \times 3 + 3 \times 4}{100} = 1.36.$$

### ★ 2.3.2 Alternatives to the Arithmetic Mean

The *geometric mean* of two numbers is the length of the side of a square of area equal to the product of the two numbers. For $N$ numbers it is defined as

$$\sqrt[N]{x_1 x_2 x_3 \cdots x_N}.$$

The *harmonic mean* is the reciprocal of the arithmetic mean of the reciprocals:

$$\frac{N}{1/x_1 + 1/x_2 + 1/x_3 + \cdots + 1/x_N}.$$

Pythagoras discovered that notes from strings whose lengths were in the ratio $1:\frac{1}{2}:\frac{1}{3}:\cdots$ were pleasing or 'harmonic'. Hence for two numbers the harmonic mean is the intermediate value such that the three reciprocals are in arithmetic progression and the numbers belong to the sequence.

The *root mean square* is just what it says, i.e.

$$\sqrt{\frac{x_1^2 + x_2^2 + x_3^2 + \cdots + x_N^2}{N}}.$$

All of these are less common than the arithmetic mean, so that if the 'mean' is mentioned without a qualifying adjective, this refers to the arithmetic mean.

The *mode* is the most popular value in a set of data. It is easy to find, but it can be a misleadingly unrepresentative number to quote.

The *median* is the halfway point, in the sense that half the data elements fall below it, and half above. It is preferable to the mean in describing data where the order or *rank* in the variable is more important than the numerical value.

Actually, although the median is generally defined as the point with half the data below and half above, it is not really quite so simple. If the data have an odd number of elements, all with different values, then the median is taken as the middle one. This therefore has $(N-1)/2$ values above it and $(N-1)/2$ below, so there are in fact slightly less than half below and above. If there are several data points with the same value, perhaps because the data have been binned, then the best one can do to define a 'central' bin is to state that not more than half lie below it, and not more than half above, i.e. there will also be some at that value—and the numbers above and below may be different. If the number of elements is even there is a further complication if the two midmost points have different values, as then any number between them would satisfy the definition; the median is, by convention, taken as halfway between the two.

## 2.4 MEASURING THE SPREAD

### 2.4.1 The Variance

The mean $\bar{x}$ describes all your data with just one number. This can be useful, but it can also be misleading. Consider the two sets of (fictitious) data in Figure 2.3.

Both sets of marks have a mean of 7.0, but they differ greatly. The first assessor departs from the average only when the student is outstandingly good or bad, and then only by a small amount, but the second marks over a wider range. They are distributed very differently, and we need a number to express the spread or *dispersion* of the data about the mean.

The average deviation from the mean is not a useful quantity, as the positive and negative deviations cancel and the sum is automatically zero.

$$\frac{1}{N}\sum_i (x_i - \bar{x}) = \frac{1}{N}\sum_i x_i - \frac{1}{N}\sum_i \bar{x}$$
$$= \bar{x} - \bar{x}$$
$$= 0.$$

However, you can stop the contributions from different elements cancelling by squaring them, which forces them to be positive. Thus the average *squared* deviation from the mean is a sensible measure of the spread of the data. It is called the *variance* of $x$ as it expresses how much $x$ is liable to vary from

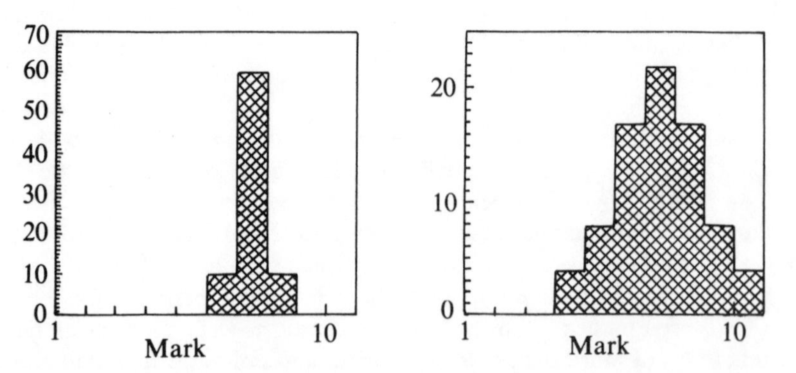


Fig. 2.3. Histograms showing the marks awarded by two demonstrators in assessing the performances of 80 students in the laboratory.

its mean value $\bar{x}$, and is written $V(x)$:

$$V(x) = \frac{1}{N}\sum_i (x_i - \bar{x})^2. \tag{2.5}$$

In the same way, any function of $x$ has a variance too:

$$V(f) = \frac{1}{N}\sum_i (f(x_i) - \bar{f})^2. \tag{2.6}$$

The definition of $V(x)$ can be manipulated to give a simpler formula for it. This type of manipulation is used so often that this time we will go through it in detail.

| | |
|---|---|
| Starting from | $V = \frac{1}{N}\sum_i (x_i - \bar{x})^2$ |
| multiply out the square | $= \frac{1}{N}\sum_i (x_i^2 - 2x_i\bar{x} + \bar{x}^2)$ |
| separate into three sums | $= \frac{1}{N}\sum_i x_i^2 - \frac{1}{N}\sum_i 2x_i\bar{x} + \frac{1}{N}\sum_i \bar{x}^2$ |
| extract some factors to get | $= \frac{1}{N}\sum_i x_i^2 - \frac{1}{N}2\bar{x}\sum_i x_i + \frac{1}{N}\bar{x}^2\sum_i 1$ |
| which reduces to | $= \overline{x^2} - 2\bar{x}^2 + \bar{x}^2$ |
| and finally | $= \overline{x^2} - \bar{x}^2.$ |

So the fundamental formula is obtained

$$V(x) = \overline{x^2} - \bar{x}^2 \tag{2.7a}$$

or, equivalently,

$$V(x) = \frac{1}{N}\sum x_i^2 - \left(\frac{1}{N}\sum x_i\right)^2 \tag{2.7b}$$

or, in words, the variance is the mean square minus the squared mean.

### 2.4.2 The Standard Deviation

The root mean squared deviation is called the *standard deviation* and given the symbol $\sigma$. It is just the square root of the variance (see previous section) and can be expressed in various equivalent forms (using equations 2.5 or 2.7):

$$\sigma = \sqrt{V(x)} \tag{2.8a}$$

$$\sigma = \sqrt{\overline{x^2} - \bar{x}^2} \tag{2.8b}$$

$$\sigma = \sqrt{\frac{1}{N}\sum_i x_i^2 - \left(\frac{1}{N}\sum_i x_i\right)^2} \tag{2.8c}$$

$$\sigma = \sqrt{\frac{1}{N}\sum_i (x_i - \bar{x})^2}. \tag{2.8d}$$

$\sigma$ represents a reasonable amount for a particular data point to differ from the mean. The exact numerical details depend on the case, but usually one is not surprised by data points one or two standard deviations from the mean, whereas a data point three or more $\sigma$ away would cause a few raised eyebrows.

Broadly speaking, practical scientists like to work with $\sigma$ rather than $V$, as it has the same units and dimensions as $x$. Statisticians, on the other hand, tend to use $V$ as it is easier to manipulate. It does not really matter which you use, as it is trivial to translate from one to the other.

*Example Laboratory marks*
The narrow histogram in Figure 2.3 has 10 cases of 6 marks, 60 of 7 marks, and 10 of 8 marks. The gives a mean of 7, a variance of 0.25, and thus a standard deviation of 0.50 marks. The broad one has a standard deviation of 1.46 marks, nearly three times larger.

*Example Monitoring*
A company produces ball-bearings whose mean mass is 30 grams, with a standard deviation of 0.1 gram. Quality control inspectors check the production line by weighing a ball-bearing every morning. If its mass lies between 29.8 and 30.2 grams they assume all is well. If it is outside these $2\sigma$ limits—the 'warning level'—but within 29.7 and 30.3 grams they weigh some more ball-bearings. If it is outside the $3\sigma$ limits of 29.7 and 30.3 grams—the 'action level'—they halt the production line.

### ★ 2.4.3 Different Definitions of the Standard Deviation

The definition of $\sigma$ is a minefield of alternatives, and to call it the 'standard' deviation is something of a sick joke. It is important to face up to this, for when people are unaware of the differences between the definitions they get confused and dismayed by factors of $\sqrt{N/(N-1)}$ that appear apparently out of nowhere. This leads to a tendency to insert such factors at random and generally incorrect moments.

Equation 2.8 defined the standard deviation of a data sample as

$$\sigma = \sqrt{\frac{1}{N}\sum_i (x_i - \bar{x})^2}.$$

So far so good. However, our data are presumably taken as a sample from a parent distribution,[†] which has a mean and a standard deviation, denoted $\mu$ and $\sigma$. In terms of expectation values:

$$\mu = \langle x \rangle$$
$$\sigma = \sqrt{\langle x^2 \rangle - \langle x \rangle^2}. \tag{2.9}$$

There is thus a clear distinction between $\bar{x}$, the mean of the sample, and $\mu$, that of the parent, and complete confusion between $\sigma$, the standard deviation of the sample, and $\sigma$, that of the parent.

This is not really too bad, as it is generally clear which is meant. However, it gets worse. Some authors define the term 'standard deviation' as the r.m.s. deviation of the data points from the 'true' mean $\mu$, rather than the sample mean $\bar{x}$:

$$\sqrt{\frac{1}{N}\sum_i (x_i - \mu)^2}. \tag{2.10}$$

This is felt to be a more fundamental and 'truer' quantity than that defined in equation 2.8, but it is not much use if you do not know the value of $\mu$. However, an estimate of this, which (when squared) gives an unbiased estimate of $\sigma^2$ of the parent, is given by

$$s = \sqrt{\frac{1}{N-1}\sum_i (x_i - \bar{x})^2}. \tag{2.11}$$

This is fair enough, and is considered in detail in Chapter 5, but we now have four definitions of 'standard deviation', three of which (equations 2.8, 2.10, and 2.11) are to some extent rivals.

The reason for all this regrettable mess is a chicken- and -egg argument as to which comes first (i.e. is more fundamental), equation 2.8 or 2.9. One school of thought says that there is a real, true, ideal distribution with a standard deviation defined by 2.9, which is best measured by 2.10 or, failing that, 2.11, and the standard deviation of your sample, as defined by 2.8, has no real significance. The opposing view, to which I incline myself, is that equation 2.8 is what you actually measure, and from a descriptive point of view that is that; any further developments towards properties of the parent distribution come under the heading of inference. After all, there are some distributions, such as the Cauchy distribution (see section 3.5.3), for which the expectation values in equation 2.9 do not exist, and yet one can perfectly well take a sample from such a distribution (nuclear physicists do it all the time) which

[†]Distributions are discussed in Chapter 3. The current section is starred material and assumes some knowledge from later chapters. This ordering is regrettably necessary to bring together all material on $\sigma$.

will have a meaningfully measurable standard deviation in the sense of equation 2.8. Anyway, it is not a matter of 'right' and 'wrong' definitions: you can use whichever definition of standard deviation you please, provided you make it clear to other people what that is, and when using other people's results and formulae involving $\sigma$ or $s$ you check what they mean by it.

Some authors helpfully use the name *sample standard deviation* explicitly for the quantity defined in equation 2.11. Unfortunately others use it for the quantity defined in equation 2.8. Definitions of variance, and *sample variance*, are similarly confused.

In this book we will consistently use $\sigma$ as defined in equation 2.8 and $s$ for the quantity defined by 2.11. This is not universal, and different authors use either symbol for either quantity—you have been warned. Some authors use Greek symbols for quantities from distributions and the Roman alphabet for those of data samples, but the usage of $\sigma$ is so entrenched that this has no chance of universal adoption, and anyway this still leaves the ambiguity between equations 2.8, 2.10, and 2.11. If necessary, the distinction can be made completely clear and explicit by denoting the quantity defined by equation 2.8 as $\sigma_N$ and that of equation 2.11 as $\sigma_{N-1}$, though this involves extra subscripts which lead to messy-looking formulae.

### ★ 2.4.4 Alternative Measures of the Spread

Another obvious way to stop the contributions from different deviations cancelling, as was done in section 2.4.1, is to take not their squares but their moduli. You can do this and form

$$\frac{1}{N}\sum_i |x_i - \bar{x}|$$

which is called the *mean absolute deviation* or sometimes, misleadingly, the *mean deviation*. This is very rarely used, because the variance and standard deviation have much more amenable mathematical behaviour—differentiating a squared quantity leads to easy, linear terms, whereas a differentiated modulus is horrible to work with.

The *range* of the data is just the difference between the highest and lowest values. It is easy to work out, but suffers from the fact that these extreme values are usually out in the tails of the data, and the value of the range is thus subject to large fluctuations, making it an unreliable number to quote.

The *interquartile range* is appropriate in the same sort of case where one would use the median, when the ordering of the data is more significant than the numerical values. The *lower quartile* is the point with 25% of the data below it and 75% above; likewise the *upper quartile* has 75% below and 25% above, and the interquartile range is the distance between them. There are also *deciles*, which contain 10% of the data, and *percentiles*, which contain a given percentage in the obvious way.

One problem with $\sigma$ is that its value can be dominated by a few extreme values out in the tails of the data. The *full width at half maximum* or *FWHM* (or full width at half height) is independent of the tails, using only the central peak. It is just what it says, and is easy to find from a histogram using a pencil and ruler: you measure the height at the peak, draw a horizontal line at half this height, and then read off the distance between the two points where the line intersects the histogram. If you have to compare widths of distributions, some with $\sigma$ quoted and others described by the FWHM, you can do so (with care!) using the fact that *for a Gaussian distribution* (see section 3.4)

$$\text{FWHM} = 2.35\sigma. \tag{2.12}$$

## ★2.5 HIGHER POWERS OF $x$

The mean and standard deviation were calculated from the first and second powers of the data values $x_i$. By obvious extensions, further numbers can be obtained from higher powers.

### ★2.5.1 Skew

The *skew* is a way of describing the asymmetry of the data. It is made from the third power of $x$, and is defined as

$$\begin{aligned}\gamma &= \frac{1}{N\sigma^3}\sum_i (x_i - \bar{x})^3 \\ &= \frac{1}{\sigma^3}\overline{(x - \bar{x})^3}.\end{aligned} \tag{2.13}$$

Expanding the definition, as was done for the variance in section 2.4.1, gives

$$\gamma = \frac{1}{\sigma^3}(\overline{x^3} - 3\bar{x}\overline{x^2} + 2\bar{x}^3). \tag{2.14}$$

The $\sigma^3$ factor in the definition makes $\gamma$ a dimensionless number. $\gamma$ is zero if the data are symmetrically distributed about the mean. If a tail extends to the right, $\gamma$ is positive and the data are *positively skew*. *Negatively skew* data have $\gamma$ negative and a tail to the left.

Skew is not used much by physical scientists, who are mainly concerned with errors of measurement. These are usually as likely to be too large as too small and so the distribution is symmetrical. The Maxwell distribution of velocities of molecules in a gas is perhaps the most familiar example of a skew distribution.

There are several alternative ways of defining skew, and accordingly one has to be careful. One common and convenient one is *Pearson's skew*, which is

$$\text{Skew} = \frac{\text{mean} - \text{mode}}{\sigma}.$$

### ★ 2.5.2 Higher Powers

*Curtosis* is made from the fourth power of $x$:

$$c = \frac{1}{\sigma^4}\overline{(x - \bar{x})^4} - 3. \tag{2.15}$$

Expanding gives

$$c = \frac{1}{\sigma^4}(\overline{x^4} - 4\bar{x}\overline{x^3} + 6\overline{x^2}\bar{x}^2 - 3\bar{x}^4) - 3 \tag{2.16}$$

Again, definitions vary, so if you meet it, be careful. Indeed, even the spelling varies—it can also be spelt *kurtosis.* It is dimensionless, thanks to the fourth power of $\sigma$ in the denominator, and for the Gaussian (normal) distribution $c$ is zero—the 3 in the definition is brought in specifically to ensure this.

Positive $c$ implies a relatively higher peak and wider wings than the Gaussian distribution with the same mean and standard deviation. Negative $c$ means a wider peak and shorter wings.

Curtosis is not used much by physicists, chemists, or indeed by anyone else. It is a really obscure and arcane quantity whose main use is inspiring awe in demonstrators, professors, or anyone else you need to impress.

In general

$$\frac{1}{N}\sum_i x_i^r \tag{2.17}$$

is called the $r$th *moment* of $x$, and

$$\frac{1}{N}\sum_i (x_i - \bar{x})^r \tag{2.18}$$

is called the $r$th *central moment* of $x$. Skew and curtosis are just (apart from a couple of unimportant constants) the third and fourth central moments.

## 2.6 MORE THAN ONE VARIABLE

In some cases each item of data consists of not just one value but two, three, or more. For example, one might record the position and time of a moving particle, so that the data are a set of pairs of $(x, t)$ measurements. Again, the height, weight, IQ, and measured physical stamina for a class of students could be recorded, so there would be one data item for each student, each made up of four individual results. This adds a new aspect to the properties of the data sample, and one can investigate the relationship between the quantities.

### 2.6.1 Covariance

Suppose each item of a data sample consists of a *pair* of numbers, $\{(x_1, y_1), (x_2, y_2), (x_3, y_3), \ldots\}$. You can find their means, $\bar{x}$ and $\bar{y}$, variances, $V(x)$ and $V(y)$, and standard deviations, $\sigma_x$ and $\sigma_y$. However, there is more information there. You can also look at the two variables together—are they independent or do they depend on one another? This is described by the *covariance* between $x$ and $y$, which is defined as

$$\operatorname{cov}(x, y) = \frac{1}{N}\sum_i (x_i - \bar{x})(y_i - \bar{y}) \tag{2.19a}$$

$$= \overline{(x - \bar{x})(y - \bar{y})} \tag{2.19b}$$

$$= \overline{xy} - \bar{x}\,\bar{y}. \tag{2.19c}$$

If values of $x$ that are above average have a tendency to occur together with above-average $y$ values (which implies that small $x$ likewise tend to accompany small $y$), then the signs of the two terms in the elements of the sum will tend to be the same, and the net sum in equation 2.19*a* will be positive. Likewise if large $x$ tend to go with small $y$, the covariance is negative. If they are independent and unconnected then a positive $x_i - \bar{x}$ has an equal chance of being multiplied by a positive or negative $y_i - \bar{y}$, and the net sum is zero.

Notice that as $\operatorname{cov}(x, y)$ involves only differences from the mean $x$ and $y$, it does not change if the origin is shifted, and that it is a generalisation of the variance, in that $\operatorname{cov}(x, x) = V(x)$.

*Example Covariances and relationships*
The covariance between height and weight in a group of adults is presumably positive, as tall people tend to weigh more. Between weight and stamina it may well be negative, as overweight people may be out of condition in other ways. Between height and IQ it is probably zero, as there is no obvious reason for tall people to be more clever, or more stupid, than short people.

### 2.6.2 Correlation

The covariance is useful, but has dimensions. A covariance between height and weight of 7.6, say, means one thing in centimetre-grams and another in metre-kilograms. A better measure of the relation between two variables is the *correlation coefficient*, $\rho$. This is defined as

$$\rho = \frac{\text{cov}(x, y)}{\sigma_x \sigma_y} \tag{2.20a}$$

$$= \frac{\overline{xy} - \bar{x}\bar{y}}{\sigma_x \sigma_y}. \tag{2.20b}$$

$\rho$ is a number between $-1$ and $+1$. If $\rho$ is zero then $x$ and $y$ are *uncorrelated.* A *positive* correlation means that if a particular $x$ happens to be larger than the mean $\bar{x}$, then $y$ will also (on average) be larger than the mean $\bar{y}$. For a negative $\rho$, a larger $x$ will imply a smaller $y$. If $\rho$ is 1 (or $-1$) then $x$ and $y$ are completely correlated: if you know the value of one that specifies precisely the value of the other. $\rho$ is dimensionless, and is unaffected by shifts in the origin or by changes in the scale for $x$ or $y$ (see Figure 2.4).

Here are some real examples of correlation, from the performances of a class of second-year students who took exams in thermal physics and quantum mechanics, and were also assessed on their laboratory work.

Figure 2.5 shows the marks obtained in quantum mechanics (horizontally) and thermal physics (vertically). As one would expect, students who do well in one tend to do well in the other, and the correlation coefficient is 0.7.

Figure 2.6 shows the marks of the same students in laboratory work, again

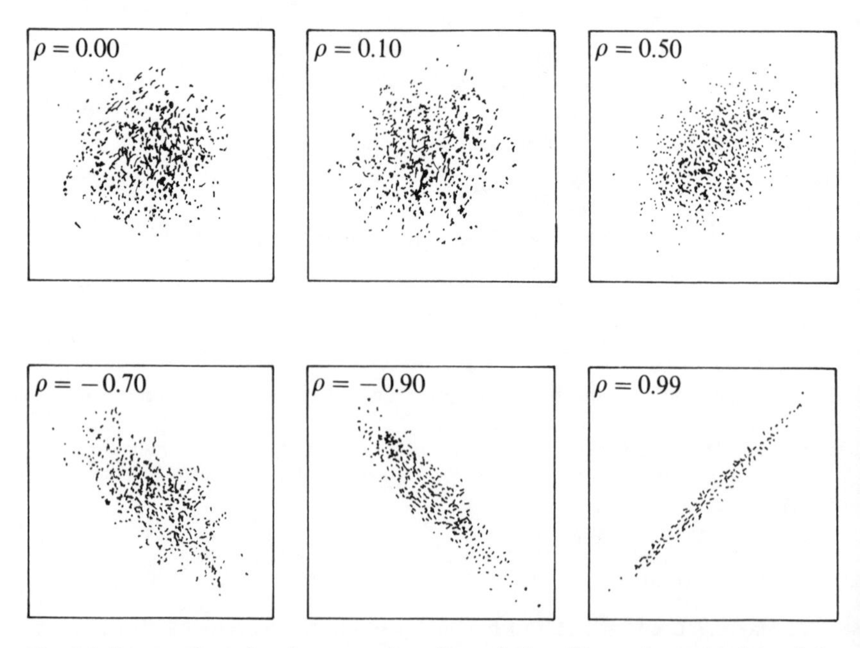


Fig. 2.4 Scatter plots showing examples of correlation. The scales and origin of the axes are irrelevant (see text) and are therefore not shown.

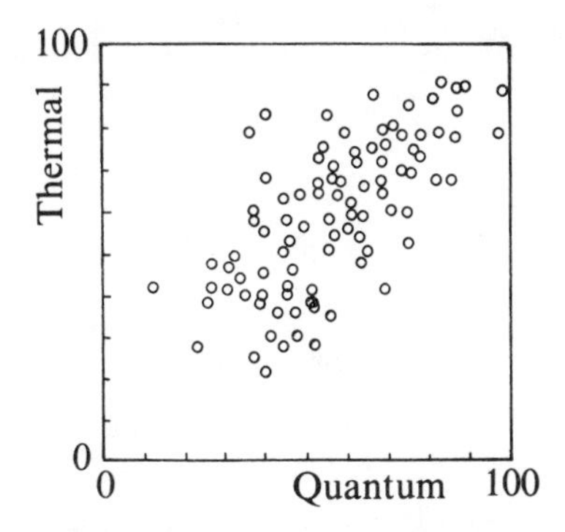


Fig. 2.5 Marks in quantum mechanics and thermal physics.

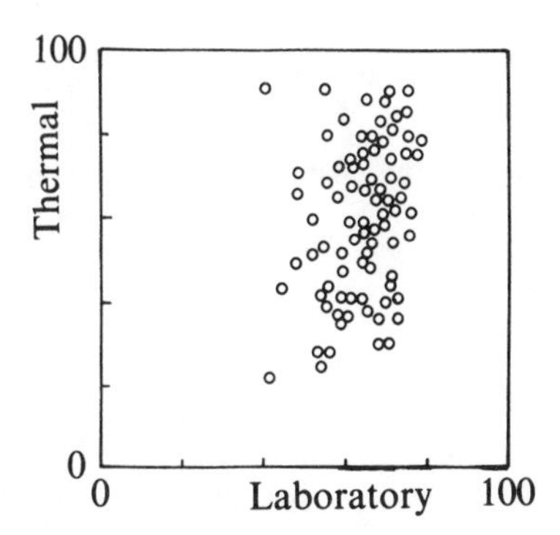


Fig. 2.6 Marks in laboratory and thermal physics.

compared with thermal physics. The tendency for students to do well or badly in both is still present, but much weaker, and $\rho$ is only 0.3.

### ★2.6.3 More Than Two Variables

If there are three elements in each item we can call them $(x, y, z)$, but owing to our lack of foresight in putting $x$ so near the end of the alphabet, this does not work for more complicated cases. Therefore, if an item comprises $n$ elements then we write it as $(x_{(1)}, x_{(2)}, x_{(3)}, \ldots, x_{(n)})$.

> Notice the notation here carefully. The bracketed subscripts denote an element within an item, and range from 1 to $n$. Thus $x_{(2)}$ is the same as $y$ above. Unbracketed subscripts denote items within the sample, as usual, and range from 1 to $N$. Of course, if you have a particular instance to work with, as opposed to the general case given here, you are free to call your data what you choose.

There is a covariance between each pair of variables, defined according to equation 2.19:

$$\operatorname{cov}(x_{(i)}, x_{(j)}) = \overline{x_{(i)}x_{(j)}} - \overline{x_{(i)}}\,\overline{x_{(j)}}. \tag{2.21}$$

With this notation you can see that these form the elements of an $n \times n$ symmetric matrix:

$$V_{ij} = \operatorname{cov}(x_{(i)}, x_{(j)}). \tag{2.22}$$

This is called the *covariance matrix*, or the *variance matrix* or the *error matrix*, and the diagonal elements are just the variances.

The *correlation matrix* is the dimensionless equivalent of the covariance matrix. Its elements must lie between $-1$ and $+1$, and give the extent to which two variables are correlated. Its diagonal elements are all 1:

$$\rho_{ij} = \frac{\operatorname{cov}(x_{(i)}, x_{(j)})}{\sigma_i \sigma_j}. \tag{2.23}$$

This gives a convenient way of writing the covariance matrix:

$$V_{ij} = \rho_{ij}\sigma_i\sigma_j. \tag{2.24}$$

## 2.7 PROBLEMS

*2.1*

The ages (in years) of a class of 25 students are as follows:

19.0, 18.7, 19.3, 19.2, 18.9, 19.0, 20.2, 19.9, 18.6, 19.4, 19.3, 18.8, 19.3, 19.2, 18.7, 18.5, 18.6, 19.7, 19.9, 20.0, 19.5, 19.4, 19.6, 20.0, 18.9

Calculate the mean and standard deviation.

*2.2*

Repeat the calculation for problem 2.1, including the lecturer (age 37.0). Notice that the effect on the mean is modest, but the standard deviation increases greatly.

★ *2.3*

Repeat problems 2.1 and 2.2, calculating the skew.

*2.4*

The marks of twelve students in classical mechanics and quantum mechanics are as follows:

| Classical | 22 | 48 | 76 | 10 | 22 | 4 | 68 | 44 | 10 | 76 | 14 | 56 |
|---|---|---|---|---|---|---|---|---|---|---|---|---|
| Quantum | 63 | 39 | 61 | 30 | 51 | 44 | 74 | 78 | 55 | 58 | 41 | 69 |

Calculate the two average marks, the covariance, and the correlation.

★ *2.5*
Prove the formulae 2.14 and 2.16.

*2.6*
Here are 80 numbers:

| | | | | | | | | | |
|---|---|---|---|---|---|---|---|---|---|
| 90 | 90 | 79 | 84 | 78 | 91 | 88 | 90 | 85 | 80 |
| 88 | 75 | 73 | 79 | 78 | 79 | 67 | 83 | 68 | 60 |
| 73 | 79 | 69 | 74 | 76 | 68 | 72 | 72 | 75 | 60 |
| 61 | 66 | 66 | 54 | 71 | 67 | 75 | 49 | 51 | 57 |
| 62 | 64 | 68 | 58 | 56 | 79 | 63 | 68 | 64 | 51 |
| 58 | 53 | 65 | 57 | 59 | 65 | 48 | 54 | 55 | 40 |
| 49 | 42 | 36 | 46 | 40 | 37 | 53 | 48 | 44 | 43 |
| 35 | 39 | 30 | 41 | 41 | 22 | 28 | 36 | 39 | 51 |

Histogram them using a suitable bin size.

★ *2.7*
Find the mean, mode, and median for the data in the previous problem, using the raw and the binned data.

★ *2.8*
Using the data in the previous question, calculate the standard deviation and the FWHM. Do they conform to equation 2.12?

★ *2.9*
Show that the magnitude of the correlation coefficient $\rho$ cannot exceed 1.

*It is a capital mistake to theorise before one has data. Insensibly one begins to twist the facts to suit theories, instead of theories to suit facts*

*—Sir Arthur Conan Doyle*

CHAPTER 

# Theoretical Distributions

Science is supposed to explain to us what is actually happening, and indeed what will happen, in the world. Unfortunately as soon as you try and do something useful with it, sordid arithmetical numbers start getting in the way and messing up the basic scientific laws. An unbiased coin may perfectly well come down heads uppermost 55 times out of 100. A decaying radioactive source may give 95 counts on a Geiger counter in one minute, and 110 counts in the next. A 10 volt power supply across a resistor marked 100 Ω may give a reading of 103 mA on your ammeter. Predictions from basic laws are modified by statistical distributions, arising from the finite size of the data sample, the experimental accuracy, and similar causes. This chapter deals with the basic ideas of distributions, and especially with the three fundamental statistical distributions: the binomial, the Poisson, and the Gaussian. Only by understanding the ways the distributions give rise to the data can one go on to use the particular behaviour of the data to produce general statements about the processes that produced them in the first place—or, as Holmes puts it, to twist your theory to suit your observed facts.

## 3.1 GENERAL PROPERTIES OF DISTRIBUTIONS

### 3.1.1 A Simple Distribution

Suppose you toss four coins. This is a simple example, and I will not pretend it is of any intrinsic interest; nevertheless we will go through it in detail as it introduces concepts that will be needed later for real problems.

For each coin the probability of the head landing uppermost is $\frac{1}{2}$, and so is the probability for the tail. We want to discuss the various possible outcomes for the four coins, and their probabilities.

1. The four coins could all land head uppermost. The probability of the first coin giving a head is $\frac{1}{2}$; so are those for the second, third, and fourth. To find the combined probability of all four giving a head we multiply the individual probabilities together, so the probability of four heads is $(\frac{1}{2})^4$. Call this $P(4)$; then $P(4) = \frac{1}{16}$.
2. Suppose the first three coins land head upwards, the fourth tail upwards. The combined probability for this is again the product of the individual ones, which gives $\frac{1}{8}$ for the first three and $\frac{1}{2}$ for the fourth, as the probability of a tail is also $\frac{1}{2}$, again giving $\frac{1}{16}$. However, if we ask for three heads and one tail, without specifying which coin gives the tail, there are four choices, namely HHHT, HHTH, HTHH, and THHH, each with the same probability of $\frac{1}{16}$, so the total probability $P(3)$ for three heads and a tail is: $P(3) = 4 \times \frac{1}{16} = \frac{1}{4}$.
3. For two heads and two tails there are six permutations of coins—HHTT, HTHT, HTTH, TTHH, THTH, and THHT—each of probability $\frac{1}{16}$, so the probability $P(2)$ of getting two heads and two tails is $\frac{3}{8}$.
4. For one head and three tails the probability is the same as one tail and three heads, so we can write down at once $P(1) = P(3) = \frac{1}{4}$.
5. Likewise, for no heads and four tails, $P(0) = P(4) = \frac{1}{16}$.

A quick check can be done by making sure that the total probability of something happening is 1:

$$\sum_r P(r) = P(0) + P(1) + P(2) + P(3) + P(4) = \frac{16}{16} = 1.$$

So if $r$ is the number of heads ($r = 0, 1, 2, 3, 4$), we have a collection of probabilities $P(r) = (\frac{1}{16}, \frac{1}{4}, \frac{3}{8}, \frac{1}{4}, \frac{1}{16})$, giving the probability that a toss of four unbiased coins will give $r$ heads. This is a simple example of a *probability distribution.*

### 3.1.2 The Law of Large Numbers

Having all these numbers, let us try and do something with them. The probabilities are in the ratio 1:4:6:4:1, i.e. if one tosses four coins sixteen times, there should be one result with four heads, four with three heads and a tail, etc. Four coins were accordingly tossed sixteen times, and the results are shown in the following table:

| Number of heads | $r=4$ | $r=3$ | $r=2$ | $r=1$ | $r=0$ |
|---|---|---|---|---|---|
| Theory predicts | 1 | 4 | 6 | 4 | 1 |
| Data | 2 | 7 | 2 | 4 | 1 |

They do not agree. There is certainly a similarity in the pattern, but the numbers do not match perfectly. Indeed, it would have been surprising if they had. With such a small number of tosses (only sixteen) statistical fluctuations are substantial. To give the numbers a chance, the experiment was repeated with a 160, 1600, and 16 000 trials:[†]

| Number of heads | $r=4$ | $r=3$ | $r=2$ | $r=1$ | $r=0$ |
|---|---|---|---|---|---|
| 160 tosses | | | | | |
| Theory predicts | 10 | 40 | 60 | 40 | 10 |
| Data | 10 | 40 | 61 | 38 | 11 |
| 1600 tosses | | | | | |
| Theory predicts | 100 | 400 | 600 | 400 | 100 |
| Data | 125 | 403 | 567 | 409 | 96 |
| 16 000 tosses | | | | | |
| Theory predicts | 1000 | 4000 | 6000 | 4000 | 1000 |
| Data | 1009 | 3946 | 5992 | 4047 | 1006 |

The agreement becomes better and better as the number of trials increases and random effects are smoothed out.

The theory predicts a set of probabilities. The observed data frequencies do not quite agree with them. However, as the size of the data sample $N$ increases the fluctuations cancel out, and the frequencies tend to the probabilities as $N$ tends to infinity. This is the *law of large numbers.*

### 3.1.3 Expectation Values

If you know the probability distribution for some number $r$—often, in an attempt to add excitement to the subject, called the number of 'successes'—

[†]Simulated on a computer, of course. You are urged to try some cointossing experiments of your own, to appreciate the way in which the experimental distributions never quite agree with the theoretical ideal.

one thing you can easily compute is the average number of 'successes' you would expect. This is called the *expectation value* of $r$ and is written $\langle r \rangle$, or sometimes $E(r)$.[†] It is given by

$$\langle r \rangle = \sum_r rP(r). \tag{3.1}$$

For example, with four coins, as discussed in section 3.1.1, the average number of heads is given by

$$0 \times \frac{1}{16} + 1 \times \frac{1}{4} + 2 \times \frac{3}{8} + 3 \times \frac{1}{4} + 4 \times \frac{1}{16} = 2$$

which is an obvious result, but shows how the formula works.

Note that $\langle r \rangle$ is not necessarily the most probable result, although it is in this example. For five coins, $\langle r \rangle = 2.5$.

Any function of $r$ also has its expectation value, defined in the same way:

$$\langle f \rangle = \sum_r f(r)P(r). \tag{3.2}$$

One useful way to think of the expectation value is in terms of gambling; suppose there is a random process (like a fruit machine) with various possible outcomes $r$, each of which has probability $P(r)$, and pays out an amount $f(r)$. Then the expectation value $\langle f \rangle$ is what you would expect, on average, to win, and would be an exactly fair fee to pay the organiser of the game for taking part.

There is an obvious parallel between an expectation value and the mean of a data sample (as described in the previous chapter). The former is a sum over a theoretical probability distribution and the latter is a (similar) sum over a real data sample. The law of large numbers ensures that if a data sample is described by a theoretical distribution, then as $N$, the size of the data sample, goes to infinity,

$$\bar{f} \rightarrow \langle f \rangle. \tag{3.3}$$

Note that expectation values add

$$\langle f + g \rangle = \sum (f + g)P(r) = \sum f P(r) + \sum gP(r) = \langle f \rangle + \langle g \rangle$$

but they *do not* multiply. In general, $\langle fg \rangle \neq \langle f \rangle \langle g \rangle$ unless $f$ and $g$ are *independent*.

### 3.1.4 Probability Density Distributions

*Continuous* variables need treating slightly differently from discrete variables. Suppose you are measuring the lengths of a large number of pieces

[†]The expectation value of $r$ itself is also often denoted by the symbol $\mu$.

of string, randomly distributed between 10 cm and 12 cm. Somebody asks you how many are 11 cm long. The answer has to be none. There will presumably be some between 10.5 and 11.5, probably a few between 10.9 and 11.1, maybe a couple between 10.99 and 11.01, but it's unlikely there will be any in the narrow range between 10.99999 and 11.00001, and if you insist that the value has to be exactly 11.00000000000... the range is so small that the probability vanishes.

However, the probability that $x$ will lie within a specified range—like 10.9 to 11.1 cm—is a finite and perfectly sensible thing to talk about, and this is described by the *probability density distribution*, $P(x)$, defined by

$$\text{Probability (result lies between } x_1 \text{ and } x_2) = \int_{x_1}^{x_2} P(x)\,\mathrm{d}x$$

or equivalently

$$P(x) = \text{limit}_{\delta x \to 0} \frac{\text{Probability (result lies between } x \text{ and } x + \delta x)}{\delta x}.$$

Probabilities are pure numbers. Probability densities, on the other hand, have dimensions, the inverse of those of the variable $x$ to which they apply.

For expectation values the same ideas apply as for the earlier probability functions, except that you get integrals instead of summations:

$$\langle x \rangle = \int_{-\infty}^{\infty} xP(x)\,\mathrm{d}x \tag{3.4}$$

$$\langle f \rangle = \int_{-\infty}^{\infty} f(x)P(x)\,\mathrm{d}x. \tag{3.5}$$

If you have done some quantum mechanics you may have met expressions like $\langle x \rangle = \int \psi^*(x) x \psi(x)\,\mathrm{d}x$. The meaning of the symbol is exactly the same; it is the expected average value of the result. It is tempting to go further and equate the quantity $\psi^*(x)\psi(x) = |\psi(x)|^2$ with the probability density $P(x)$, but this is wrong, as it does not work for expectation values of quantities (like momentum) that involve differential operators.

## 3.2 THE BINOMIAL DISTRIBUTION

The binomial distribution describes processes with a given number of identical trials, with two possible outcomes. Examples are tossing coins (heads or tails), quality checks of components (pass or fail), treatment of patients (kill or cure), and many similar. We call the two outcomes, without prejudice, 'success' and 'failure', and denote the probability of a success as $p$, and that of failure therefore $1-p$.[†] This basic process is repeated $n$ times—$n$ is called the number of *trials*—and the distribution gives the probability of $r$ successes (and thus $n-r$ failures) out of these $n$ trials, each of which has an individual probability of success $p$.

### 3.2.1 The Binomial Probability Distribution Formula

The probability of $r$ successes from $n$ trials is a generalisation of the particular case considered in detail in section 3.1.1. It is made up of two factors. Firstly, there are $2^n$ possible permutations of success and failure, of which the number with $r$ successes is the number of ways of selecting $r$ from $n$:

$$ {}_nC_r = \frac{n!}{r!(n-r!)}. $$

Secondly, as there are $r$ successes of probability $p$, and likewise $n-r$ failures of probability $1-p$, the combined result has a probability obtained by multiplying all these together, namely $p^r(1-p)^{n-r}$

Putting these two factors together gives the

**Binomial probability distribution** *The probability of r successes out of n tries, each of which has probability p of success, is*

$$ P(r; p, n) = p^r(1-p)^{n-r}\frac{n!}{r!(n-r)!}. \tag{3.6} $$

As this probability depends not only on $r$, the number of successes, but also on the intrinsic probability $p$ and number of trials $n$, they are also shown as arguments of $P$, separated from $r$ by a semicolon. This is a purely artistic device to show that usually one considers how $P$ varies with $r$ for a given $n$ and $p$.

The ${}_nC_r$ are the binomial coefficients, so the total probability of *something* happening is the binomial expansion of $[p+(1-p)]^n$, and is therefore 1, as it has to be

$$ \sum_{r=0}^{n} p^r(1-p)^{n-r}{}_nC_r = [p+(1-p)]^n = 1^n = 1. \tag{3.7} $$

The important properties of the binomial distribution are (proofs, if desired, are given in section 3.2.2)

$$ \text{the mean number of success is} \qquad \langle r \rangle = np \tag{3.8} $$

[†]Some people define the probability of a failure as $q$. This makes formulae simpler, at the price of a new symbol and having to remember that $q$ is always equal to $1-p$. Follow your own taste in this.

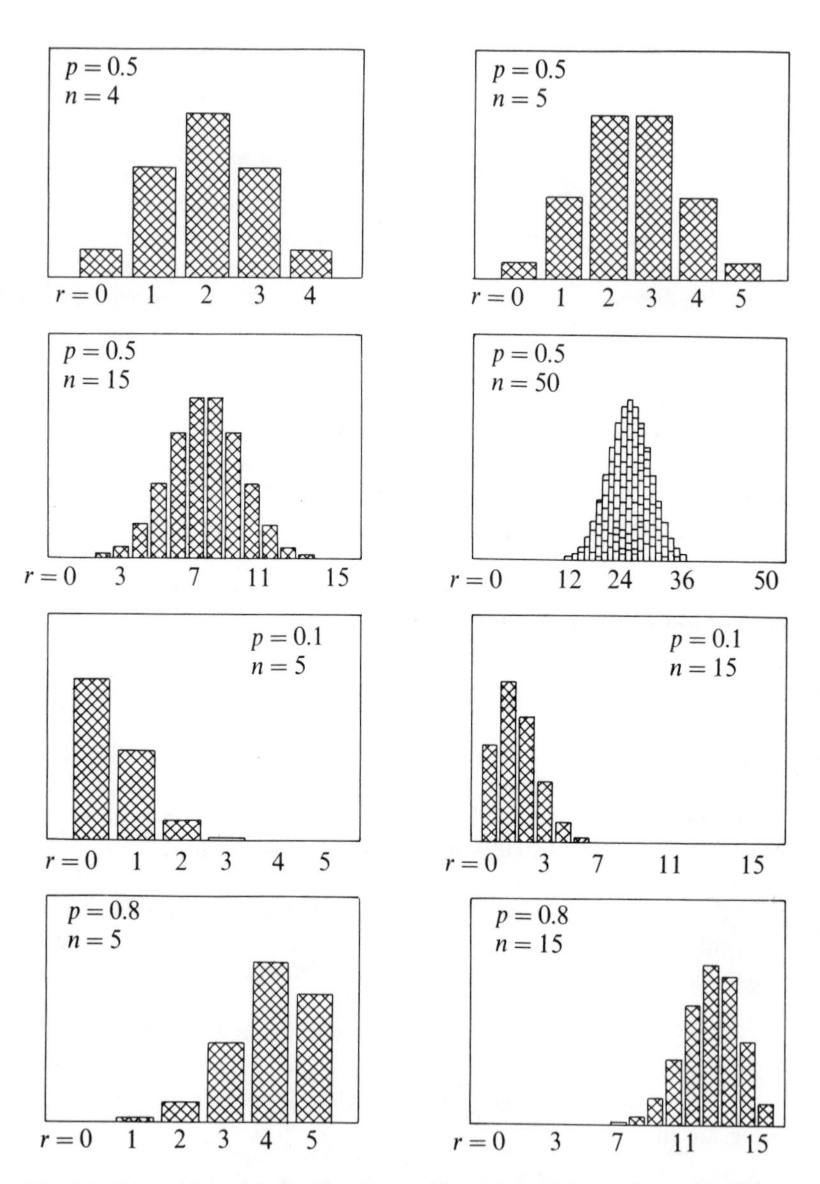


Fig. 3.1. Some binomial distributions, with various values of $n$ and $p$. (The vertical scale is arbitrary).

the variance is $V(r) = np(1-p)$ (3.9)

and thus the standard deviation is $\sigma = \sqrt{np(1-p)}$. (3.10)

Some examples of binomial distributions are shown in Figure 3.1. They peak around the value $np$, as expected. As $n$ increases, the peak, in proportion to the full range of $n$, becomes progressively narrower, albeit slowly. The relative width of the peak also depends on $p$, and (for the same $n$) peaks with $p$ close to 0 or 1 are narrower than those with $p$ near 0.5.

*Example Detector efficiencies*

You are trying to measure the tracks of cosmic ray particles using spark chambers, which are 95% efficient. You make the sensible decision that at least three points are needed to define a track. How efficient at detecting tracks would a stack of three chambers be? Would using four or five chambers give a significant improvement?

The probability of three hits from three chambers is

$$P(3; 0.95, 3) = 0.95^3 = 0.857$$

so this would be 85.7% efficient. For four chambers the probability of three or four hits is

$$P(3; 0.95, 4) + P(4; 0.95, 4) = 0.171 + 0.815 = 98.6\%.$$

For five chambers,

$$P(3; 0.95, 5) + P(4; 0.95, 5) + P(5; 0.95, 5) = 0.021 + 0.204 + 0.774 = 99.9\%.$$

*Example Guessing cards*

In an experiment into extrasensory perception, a subject guesses the symbol on a card. There are five different symbols so they have a 20% chance of guessing right by chance. If they guess six cards, what is the probability of getting more than half correct by chance?

The probability is

$$P(4; 0.2, 6) + P(5; 0.2, 6) + P(6; 0.2, 6) = 1.54\% + 0.154\% + 0.0064\% = 1.7\%.$$

### ★ 3.2.2 Proof of Properties of the Binomial Distribution

To prove equation 3.8, put the binomial formula (equation 3.6) in the expectation value (equation 3.1)

$$\langle r \rangle = \sum_{r=0}^{r=n} r p^r (1-p)^{n-r} \frac{n!}{r!(n-r)!}.$$

Take out a factor of $np$ and drop the $r = 0$ term (which is zero):

$$\langle r \rangle = np \sum_{r=1}^{r=n} p^{r-1} (1-p)^{n-r} \frac{(n-1)!}{(r-1)!(n-r)!}.$$

Substituting $r' = r - 1$, $n' = n - 1$, this becomes

$$\langle r \rangle = np \sum_{r'=0}^{r'=n'} p^{r'}(1-p)^{n'-r'} \frac{n'!}{r'!(n'-r')!}.$$

The sum is the expansion of $[p + (1-p)]^{n'}$, and is just 1 (by equation 3.7). Therefore,

$$\langle r \rangle = np.$$

To find $V(r)$, start with the expression

$$\langle r(r-1) \rangle = \sum_{r=0}^{r=n} r(r-1)p^r(1-p)^{n-r} \frac{n!}{r!(n-r)!}.$$

Similar treatment (the first two terms are now zero) gives

$$\langle r(r-1) \rangle = p^2 n(n-1) \sum_{r'=0}^{r'=n'} p^{r'}(1-p)^{n'-r'} \frac{n'!}{r'!(n'-r')!}$$

where $r' = r - 2, n' = n - 2$. The sum is again 1, so

$$\langle r^2 - r \rangle = n(n-1)p^2$$

and using $\langle r \rangle = np$

$$\langle r^2 \rangle - \langle r \rangle^2 = n(n-1)p^2 + np - (np)^2$$

$$V(r) = np(1-p)$$

which is equation 3.9.

## 3.3 THE POISSON DISTRIBUTION

The binomial distribution describes cases where particular outcomes occur in a certain number of trials, $n$. The Poisson distribution describes cases where there are still particular outcomes but no idea of the number of trials; instead these are *sharp events occurring in a continuum.* For example, during a thunderstorm there will be a definite number of flashes of lightning (sharp events), but it is meaningless to ask how often there was not a flash. A Geiger counter placed near a radioactive source will produce definite clicks, but not definite non-clicks.

If in such an experiment one knows that the average number of events is, say, ten a minute, then in a particular minute one expects on average ten events, though intuitively one feels that nine or eleven would be unremarkable... but suppose there were five or fifteen? Is that compatible, or has something changed? We need to know the probability of obtaining a particular number of events, given the average number. This can be analysed by taking the limit of the binomial distribution, in which the number of tries,

$n$, becomes large while at the same time the probability $p$ becomes small, with their product constant.

### 3.3.1 The Poisson Probability Formula

Suppose that on average $\lambda$ events would be expected to occur in some interval. Split the interval up into $n$ very small equal sections, so small that the chance of getting two events in one section can be discounted. Then the probability that a given section contains an event is $p=\lambda/n$.

The probability that there will be $r$ events in the $n$ sections of the interval is given by the binomial formula (equation 3.6)

$$P(r;\lambda/n,n)=\frac{\lambda^r}{n^r}\left(1-\frac{\lambda}{n}\right)^{n-r}\frac{n!}{r!(n-r)!}.$$

As $n\to\infty$ with $r$ finite the factorials give a power of $n$:

$$\frac{n!}{(n-r)!}=n(n-1)(n-2)\cdots(n-r+1)\to n^r$$

and an exponential appears:

$$\left(1-\frac{\lambda}{n}\right)^{n-r}\to\left(1-\frac{\lambda}{n}\right)^{n}\to e^{-\lambda}$$

(This limit is actually a definition of $e^x$; alternatively it can be seen by taking logarithms of both sides and using $\ln(1+\delta)\approx\delta$.)

Inserting these two limits in the binomial formula above gives the

**Poisson probability formula** *The probability of obtaining r events if the mean expected number is $\lambda$ is*

$$P(r;\lambda)=\frac{e^{-\lambda}\lambda^r}{r!}. \tag{3.11}$$

In calculating a series of Poisson probabilities it is often convenient to start with $P(0)$, which is just $e^{-\lambda}$, and then successively multiply by $\lambda$ and divide by $1,2,3,4,\ldots$ to get $P(1)$, $P(2)$, $P(3)$, $P(4),\ldots$.

*Example Fatal horse kicks*

The classic example of Poisson statistics is the set of figures on the numbers of Prussian soldiers kicked to death by horses. In ten different army corps, over twenty years (in the last century), there were 122 deaths, so that $\lambda$, the mean number of deaths in one corps in one year, is $\frac{122}{200}=0.610$. The probability of no deaths occurring, in a given corps for a given year, is $P(0;0.61)=e^{-0.61}0.61^0/0!=0.5434$; to get the prediction for the number of fatalities we just multiply by the number of cases considered (200) to get 108.7. Actually there were 109, so the agreement is virtually perfect. The full data show similar excellent agreement.

| Number of deaths in 1 corps in 1 year | Actual number of such cases | Poisson prediction |
|---|---|---|
| 0 | 109 | 108.7 |
| 1 | 65 | 66.3 |
| 2 | 22 | 20.2 |
| 3 | 3 | 4.1 |
| 4 | 1 | 0.6 |

*Example Supernova neutrinos*
Here are the numbers of neutrino events detected in 10 second intervals by the Irvine–Michigan–Brookhaven experiment on 23 February 1987—around which time the supernova S1987a was first seen by astronomers:

| No. of events | 0 | 1 | 2 | 3 | 4 | 5 | 6 | 7 | 8 | 9 |
|---|---|---|---|---|---|---|---|---|---|---|
| No. of intervals | 1042 | 860 | 307 | 78 | 15 | 3 | 0 | 0 | 0 | 1 |
| Prediction | 1064 | 823 | 318 | 82 | 16 | 2 | 0.3 | 0.03 | 0.003 | 0.0003 |

Ignoring the interval with nine events (for a strict justification of this see problem 8.2) the mean $\bar{r}$ is

$$\frac{860 + 307 \times 2 + 78 \times 3 + 15 \times 4 + 3 \times 5}{1042 + 860 + 307 + 78 + 15 + 3} = 0.77.$$

The Poisson predictions this gives are shown, and agree well with the data, except for the interval with nine events. This shows that the background due to random events is Poisson, and well understood, and the nine events are not a fluctuation on background, and came from the supernova.

Looking at the formula, or at the distributions shown in Figure 3.2, you can see that for $\lambda$ below 1.0, the most probable result is zero. For higher values a peak develops, but note that this is below $\lambda$—although $\lambda$ is the mean, it is not the mode. Indeed the formula shows that, for $\lambda$ integer, $r = \lambda$ and $r = \lambda - 1$ are equally probable.

The Poisson distribution is always broader than a binomial distribution with the same mean. The Poisson variance is equal to the mean $\lambda$, whereas the binomial variance $np(1-p)$ is always smaller than the mean $np$. This is understandable, as the number of binomial success does have an upper limit (as $r$ cannot exceed $n$) whereas the Poisson distribution can have a long tail. This upper tail is a characteristic of the Poisson distribution.

The important properties of the Poisson distribution (proofs, if desired, are in section 3.3.2) are

the total probability is 1 $$\sum_{r=0}^{\infty} P(r; \lambda) = 1 \qquad (3.12)$$

the mean number of events is $$\langle r \rangle = \lambda \qquad (3.13)$$

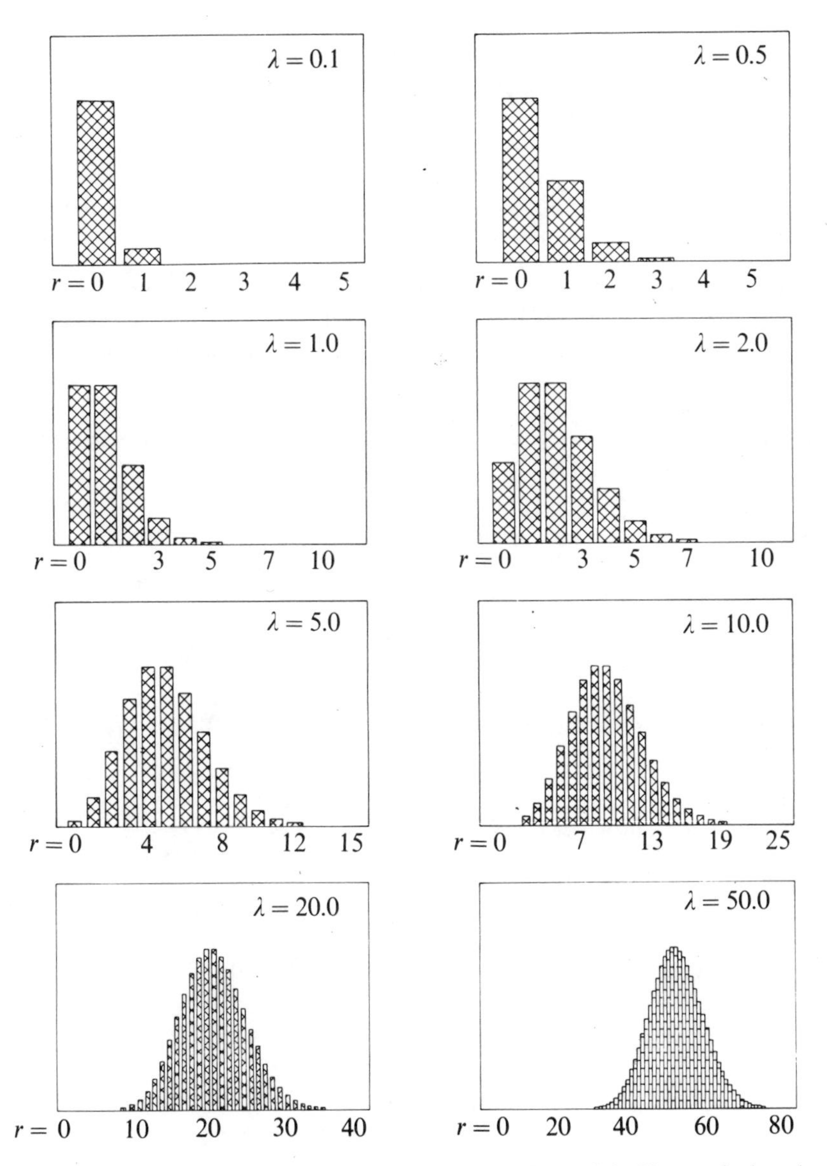


Fig. 3.2. Some Poisson distributions, with various values of $\lambda$. The vertical scale is arbitrary.

with variance $$V(r) = \lambda \tag{3.14}$$

and thus a standard deviation $$\sigma = \sqrt{\lambda} \tag{3.15}$$

and this last is overwhelmingly the most important thing to remember: for a Poisson distribution, the standard deviation is just the square root of the mean number of events.

*Example More horse kicks*
In the previous example of the Prussian horsemen, the mean was found to be 0.610. The variance is 0.608—almost identical.

The Poisson can make a useful approximation to the binomial distribution in cases where the number of trials, $n$, is large, and/or the probability $p$ is small—it is easier to calculate as it does not involve messy factorials.

*Example Poisson approximation of a binomial*
If there are 100 trials, with individual probability of success of 2%, then the binomial probabilities for the numbers of successes are

| $r$ | 0 | 1 | 2 | 3 | 4 | 5 | 6 |
|---|---|---|---|---|---|---|---|
| $P$(binomial) | 13.3% | 27.1% | 27.3% | 18.2% | 9.0% | 3.5% | 1.1% |

The Poisson distribution, for a mean of 2, gives the probabilities

| $P$(Poisson) | 13.5% | 27.1% | 27.1% | 18.0% | 9.0% | 3.6% | 1.2%. |
|---|---|---|---|---|---|---|---|

Unless you are very demanding, this accuracy is presumably ample, and the computation is much easier—try them yourself and see.

## ★ 3.3.2 Proof of Properties of the Poisson Distribution

To show that the normalisation (equation 3.12) is correct is straightforward

$$\sum_{r=0}^{\infty} P(r;\lambda) = e^{-\lambda} \sum \frac{\lambda^r}{r!}$$

$$= e^{-\lambda} e^{\lambda} \quad \text{(as the sum is just the expansion of } e^{\lambda}\text{)}$$

$$= 1.$$

$\langle r \rangle$ is given by

$$\langle r \rangle = \sum_{r=0}^{\infty} r e^{-\lambda} \frac{\lambda^r}{r!}.$$

Drop the $r=0$ term and take out some factors:

$$\langle r\rangle = \lambda \mathrm{e}^{-\lambda} \sum_{r=1}^{\infty} \frac{\lambda^{r-1}}{(r-1)!}.$$

Set $r' = r-1$:

$$\langle r\rangle = \lambda \mathrm{e}^{-\lambda} \sum_{r'=0}^{\infty} \frac{\lambda r'}{r'!}$$

and use equation 3.12 to get equation 3.13:

$$\langle r\rangle = \lambda$$

To find $V(r)$, start with

$$\langle r(r-1)\rangle = \sum_{r=0}^{\infty} r(r-1)\mathrm{e}^{-\lambda}\frac{\lambda^r}{r!}.$$

As before, dropping the first two terms and putting $r' = r-2$,

$$\langle r^2 - r\rangle = \lambda^2 \mathrm{e}^{-\lambda} \sum_{r'=0}^{\infty} \frac{\lambda^{r'}}{r'!}$$

$$\langle r^2\rangle - \langle r\rangle = \lambda^2$$

and then using equation 3.13 gives equation 3.14:

$$\langle r^2\rangle - \langle r\rangle^2 = \lambda^2 + \lambda - \lambda^2$$

$$V(r) = \lambda.$$

### ★ 3.3.3 Two Poisson Distributions

If there are two separate types of events occurring according to Poisson statistics and we do not distinguish between the two (for example, a radioactive source containing two different unstable isotopes both giving identical clicks on a Geiger counter), then the probability of $r$ events is also Poisson, with mean equal to the sum of the two means.

Suppose the two events types are called $a$ and $b$, with individual means $\lambda_a$ and $\lambda_b$, so we know the probability of observing $r_a$ and $r_b$. A total of $r$ events could be all of type $b$, or one of type $a$ and the rest of type $b$, and so on. The total probability is given by

$$\begin{aligned} P(r) &= \sum_{r_a=0}^{r} P(r_a;\lambda_a)P(r-r_a;\lambda_b) \\ &= \mathrm{e}^{-\lambda_a}\mathrm{e}^{-\lambda_b}\sum \frac{\lambda_a^{r_a}\lambda_b^{r-r_a}}{r_a!(r-r_a)!} \\ &= \mathrm{e}^{-(\lambda_a+\lambda_b)}\frac{(\lambda_a+\lambda_b)^r}{r!}\sum_{r_a=0}^{r}\frac{r!}{r_a!(r-r_a)!}\left(\frac{\lambda_a}{\lambda_a+\lambda_b}\right)^{r_a}\left(\frac{\lambda_b}{\lambda_a+\lambda_b}\right)^{r-r_a}. \end{aligned}$$

The summation, on closer inspection, is just the binomial expansion of $\left(\frac{\lambda_a}{\lambda_a+\lambda_b}+\frac{\lambda_b}{\lambda_a+\lambda_b}\right)^r$, which is just 1, so the result is

$$P(r) = e^{-(\lambda_a+\lambda_b)}\frac{(\lambda_a+\lambda_b)^r}{r!} \tag{3.16}$$

i.e. the sum of two Poisson processes is another Poisson process. This can be extended to any number of Poisson processes. The proof also shows (from the fact that the sum is a binomial expansion) that given $r$ events, the distribution of events of type $a$ is described by a binomial, $P\left(r_a; \frac{\lambda_a}{\lambda_a+\lambda_b}, r\right)$.

## 3.4 THE GAUSSIAN DISTRIBUTION

### 3.4.1 The Gaussian Probability Distribution Function

$$P(x; \mu, \sigma) = \frac{1}{\sigma\sqrt{2\pi}} e^{-(x-\mu)^2/2\sigma^2}. \tag{3.17}$$

The *Gaussian* or *normal* is the most well known and useful of all distributions. It is a bell-shaped curve centred on, and symmetric about, $x = \mu$. The width is controlled by the parameter $\sigma$, which is also the standard deviation of the distribution (which will be shown in section 3.4.2). It is broad if $\sigma$ is large, narrow if $\sigma$ is small. At $x = \mu \pm \sigma$, $P(x)$ falls to 0.61 of its peak value—at a bit more than half. These are also the points of inflexion, where the second derivative is zero.

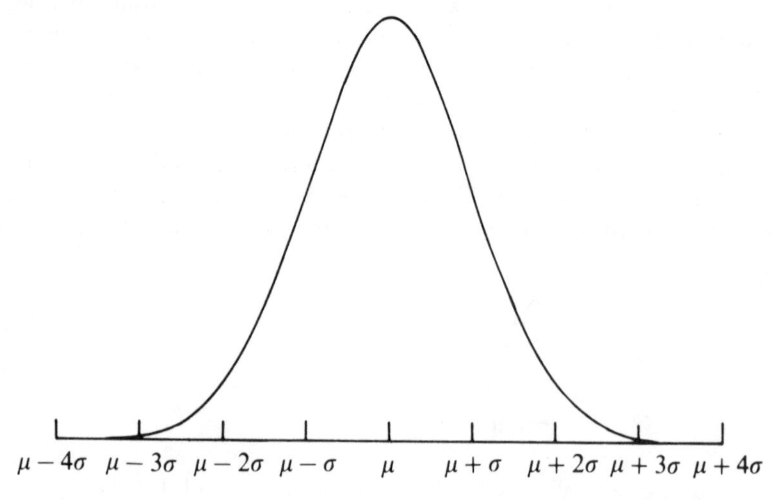


Fig. 3.3. The Gaussian distribution.

Changing the value of $\mu$ shifts the distribution along the $x$ axis without any change to its shape. Increasing or decreasing $\sigma$ stretches or shrinks the curve about the central value. In this way all Gaussians are equivalent, in that a change of origin and scale reduces them to a standard form. This is why only one Gaussian is shown here, in contrast to the many pictures of different binomial and Poisson distributions. If you substitute $z = (x - \mu)/\sigma$ then the Gaussian becomes

$$\frac{1}{\sqrt{2\pi}} e^{-z^2/2} \tag{3.18}$$

which is often called the *unit Gaussian* or *unit normal* distribution.

The important properties of the distribution (proofs, if required, are in section 3.4.2) are

it is normalized to 1:

$$\int_{-\infty}^{\infty} P(x; \mu, \sigma)\, \mathrm{d}x = 1 \tag{3.19}$$

$\mu$ is the mean of the distribution:

$$\int_{-\infty}^{\infty} xP(x; \mu, \sigma)\, \mathrm{d}x = \mu \tag{3.20}$$

(it is also the mode and the median.)
the standard deviation is $\sigma$, and variance $\sigma^2$:

$$\int_{-\infty}^{\infty} (x - \mu)^2 P(x; \mu, \sigma)\, \mathrm{d}x = \sigma^2. \tag{3.21}$$

(This justifies our use of $\sigma$ to represent both of the two quantities, the standard deviation of the distribution and the parameter in the Gaussian distribution formula, as they turn out to be the same.)

Although called after Gauss, the distribution was in fact discovered and investigated independently by many people. In France it is known as the *Laplacean.* The first recorded reference to it is by de Moivre (who was English) in 1733, in a work entitled *Approximatio ad summam terminorum binomii $(a + b)^n$ in seriem expansi.*

It is also often called the *normal* distribution. However, the use of this name implies a value-judgement (nobody, after all, would use an abnormal distribution) which is best avoided. It does indeed describe many different sorts of data, particularly in the field of measurement errors, but the reasons for this are complex and not to be glossed over by a bland label—this is the point of Lippman's famous remark (quoted by Poincaré): 'everybody believes in the law of errors, the experimenters because they think it is a mathematical theorem, the mathematicians because they think it is an experimental fact.'

### ★ 3.4.2 Proof of Properties of the Gaussian

When working with Gaussians, it is usually simpler to shift the origin so that $\mu = 0$, but to leave in the scale factor of $\sigma$, as then the dimensions make sense. To prove the normalisation, we have to show that

$$\int_{-\infty}^{\infty} \frac{1}{\sigma\sqrt{2\pi}} e^{-(x-\mu)^2/2\sigma^2} \, dx = 1.$$

Setting $x' = x - \mu$ the expression becomes

$$\frac{1}{\sigma\sqrt{2\pi}} \int_{-\infty}^{\infty} e^{-x'^2/2\sigma^2} \, dx'$$

and this integral is given in Table 3.1 (with $a = 1/2\sigma^2$), giving

$$\frac{1}{\sigma\sqrt{2\pi}} \sqrt{2\sigma^2\pi}$$

which is 1, as desired.

That $\mu$ is the mean of the distribution, which is also the expectation value $\langle x \rangle$, is obvious, but a proof can be spelt out if desired by writing

$$\langle x \rangle = \int_{-\infty}^{\infty} \frac{1}{\sigma\sqrt{2\pi}} x e^{-(x-\mu)^2/2\sigma^2} \, dx.$$

Putting $x = (x - \mu) + \mu$ and splitting the integral into two gives

$$\begin{aligned} \langle x \rangle &= \int_{-\infty}^{\infty} \frac{1}{\sigma\sqrt{2\pi}} (x-\mu) e^{-(x-\mu)^2/2\sigma^2} \, dx + \mu \int_{-\infty}^{\infty} \frac{1}{\sigma\sqrt{2\pi}} e^{-(x-\mu)^2/2\sigma^2} \, dx \\ &= 0 + \mu \times 1 \\ &= \mu. \end{aligned}$$

The variance is found from another standard integral from Table 3.1:

$$\begin{aligned} \langle (x-\mu)^2 \rangle &= \int_{-\infty}^{\infty} \frac{1}{\sigma\sqrt{2\pi}} (x-\mu)^2 e^{-(x-\mu)^2/2\sigma^2} \, dx \\ &= \sigma^2. \end{aligned}$$

### 3.4.3 Definite Integrals

In working with the Gaussian function there are various standard integrals that occur frequently. Their derivation is usually straightforward, and can be found in any reputable mathematics textbook. They are collected here for convenience.

TABLE 3.1
USEFUL INTEGRALS

$$\int_{-\infty}^{\infty} e^{-ax^2}\,dx = \sqrt{\frac{\pi}{a}} \qquad \int_{-\infty}^{\infty} e^{-z^2/2}\,dz = \sqrt{2\pi}$$

$$\int_{0}^{\infty} x\,e^{-ax^2}\,dx = \frac{1}{2a} \qquad \int_{0}^{\infty} z\,e^{-z^2/2}\,dz = 1$$

$$\int_{-\infty}^{\infty} x^2 e^{-ax^2}\,dx = \frac{1}{2a}\sqrt{\frac{\pi}{a}} \qquad \int_{-\infty}^{\infty} z^2\,e^{-z^2/2}\,dz = \sqrt{2\pi}.$$

Higher powers can be obtained by differentiating these with respect to $a$, giving

$$\int_{0}^{\infty} x^{2n+1}\,e^{-ax^2}\,dx = \frac{n!}{2a^{n+1}} \qquad \int_{0}^{\infty} z^{2n+1}\,e^{-z^2/2}\,dz = 2^n n!$$

$$\int_{-\infty}^{\infty} x^{2n}\,e^{-ax^2}\,dx = \frac{1.3.5...(2n-1)}{2^n a^n}\sqrt{\frac{\pi}{a}}$$

$$\int_{-\infty}^{\infty} z^{2n}\,e^{-z^2/2}\,dz = 1.3.5...(2n-1)\sqrt{2\pi}.$$

For any odd power, the symmetric integral vanishes:

$$\int_{-\infty}^{\infty} x^{2n+1}\,e^{-ax^2}\,dx = \int_{-\infty}^{\infty} z^{2n+1}\,e^{-z^2/2}\,dz = 0.$$

### 3.4.4 Indefinite Integrals

Unfortunately the indefinite integral of the Gaussian cannot be done analytically and written down as a nice expression. Instead you have to look it up in tables, or most reputable computers will provide a library function to evaluate it. Table 3.2 thus shows the value of the integrated Gaussian distribution, between the symmetric limits $-(x-\mu)/\sigma$ and $+(x-\mu)/\sigma$, i.e. the probability that, if an event is drawn from a Gaussian distribution, it will lie within some number of standard deviations of the mean. The probability that it will lie *outside* the range specified is, of course, just one minus the tabulated value.

From Table 3.2 you can see that

68.27% of the area lies within $\sigma$ of the mean,
95.45% lies within $2\sigma$,
99.73% lies within $3\sigma$.

If round numbers in the percentages are required, then

90% lie within $1.645\sigma$,
95% lie within $1.960\sigma$,
99% lie within $2.576\sigma$,
99.9% lie within $3.290\sigma$.

TABLE 3.2
TWO-TAILED GAUSSIAN INTEGRAL
Giving the percentage probability that a point lies within the given number of $\sigma$ from the mean

| $\frac{x-\mu}{\sigma}$ | 0.00 | 0.01 | 0.02 | 0.03 | 0.04 | 0.05 | 0.06 | 0.07 | 0.08 | 0.09 |
|---|---|---|---|---|---|---|---|---|---|---|
| 0.00 | 0.00 | 0.80 | 1.60 | 2.39 | 3.19 | 3.99 | 4.78 | 5.58 | 6.38 | 7.17 |
| 0.10 | 7.97 | 8.76 | 9.55 | 10.34 | 11.13 | 11.92 | 12.71 | 13.50 | 14.28 | 15.07 |
| 0.20 | 15.85 | 16.63 | 17.41 | 18.19 | 18.97 | 19.74 | 20.51 | 21.28 | 22.05 | 22.82 |
| 0.30 | 23.58 | 24.34 | 25.10 | 25.86 | 26.61 | 27.37 | 28.12 | 28.86 | 29.61 | 30.35 |
| 0.40 | 31.08 | 31.82 | 32.55 | 33.28 | 34.01 | 34.73 | 35.45 | 36.16 | 36.88 | 37.59 |
| 0.50 | 38.29 | 38.99 | 39.69 | 40.39 | 41.08 | 41.77 | 42.45 | 43.13 | 43.81 | 44.48 |
| 0.60 | 45.15 | 45.81 | 46.47 | 47.13 | 47.78 | 48.43 | 49.07 | 49.71 | 50.35 | 50.98 |
| 0.70 | 51.61 | 52.23 | 52.85 | 53.46 | 54.07 | 54.67 | 55.27 | 55.87 | 56.46 | 57.05 |
| 0.80 | 57.63 | 58.21 | 58.78 | 59.35 | 59.91 | 60.47 | 61.02 | 61.57 | 62.11 | 62.65 |
| 0.90 | 63.19 | 63.72 | 64.24 | 64.76 | 65.28 | 65.79 | 66.29 | 66.80 | 67.29 | 67.78 |
| 1.00 | 68.27 | 68.75 | 69.23 | 69.70 | 70.17 | 70.63 | 71.09 | 71.54 | 71.99 | 72.43 |
| 1.10 | 72.87 | 73.30 | 73.73 | 74.15 | 74.57 | 74.99 | 75.40 | 75.80 | 76.20 | 76.60 |
| 1.20 | 76.99 | 77.37 | 77.75 | 78.13 | 78.50 | 78.87 | 79.23 | 79.59 | 79.95 | 80.29 |
| 1.30 | 80.64 | 80.98 | 81.32 | 81.65 | 81.98 | 82.30 | 82.62 | 82.93 | 83.24 | 83.55 |
| 1.40 | 83.85 | 84.15 | 84.44 | 84.73 | 85.01 | 85.29 | 85.57 | 85.84 | 86.11 | 86.38 |
| 1.50 | 86.64 | 86.90 | 87.15 | 87.40 | 87.64 | 87.89 | 88.12 | 88.36 | 88.59 | 88.82 |
| 1.60 | 89.04 | 89.26 | 89.48 | 89.69 | 89.90 | 90.11 | 90.31 | 90.51 | 90.70 | 90.90 |
| 1.70 | 91.09 | 91.27 | 91.46 | 91.64 | 91.81 | 91.99 | 92.16 | 92.33 | 92.49 | 92.65 |
| 1.80 | 92.81 | 92.97 | 93.12 | 93.28 | 93.42 | 93.57 | 97.71 | 93.85 | 93.99 | 94.12 |
| 1.90 | 94.26 | 94.39 | 94.51 | 94.64 | 94.76 | 94.88 | 95.00 | 95.12 | 95.23 | 95.34 |
| 2.00 | 95.45 | 95.56 | 95.66 | 95.76 | 95.86 | 95.96 | 96.06 | 96.15 | 96.25 | 96.34 |
| 2.10 | 96.43 | 96.51 | 96.60 | 96.68 | 96.76 | 96.84 | 96.92 | 97.00 | 97.07 | 97.15 |
| 2.20 | 97.22 | 97.29 | 97.36 | 97.43 | 97.49 | 97.56 | 97.62 | 97.68 | 97.74 | 97.80 |
| 2.30 | 97.86 | 97.91 | 97.97 | 98.02 | 98.07 | 98.12 | 98.17 | 98.22 | 98.27 | 98.32 |
| 2.40 | 98.36 | 98.40 | 98.45 | 98.49 | 98.53 | 98.57 | 98.61 | 98.65 | 98.69 | 98.72 |
| 2.50 | 98.76 | 98.79 | 98.83 | 98.86 | 98.89 | 98.92 | 98.95 | 98.98 | 99.01 | 99.04 |
| 2.60 | 99.07 | 99.09 | 99.12 | 99.15 | 99.17 | 99.20 | 99.22 | 99.24 | 99.26 | 99.29 |
| 2.70 | 99.31 | 99.33 | 99.35 | 99.37 | 99.39 | 99.40 | 99.42 | 99.44 | 99.46 | 99.47 |
| 2.80 | 99.49 | 99.50 | 99.52 | 99.53 | 99.55 | 99.56 | 99.58 | 99.59 | 99.60 | 99.61 |
| 2.90 | 99.63 | 99.64 | 99.65 | 99.66 | 99.67 | 99.68 | 99.69 | 99.70 | 99.71 | 99.72 |
| 3.00 | 99.73 | 99.74 | 99.75 | 99.76 | 99.76 | 99.77 | 99.78 | 99.79 | 99.79 | 99.80 |
| 3.10 | 99.81 | 99.81 | 99.82 | 99.83 | 99.83 | 99.84 | 99.84 | 99.85 | 99.85 | 99.86 |
| 3.20 | 99.86 | 99.87 | 99.87 | 99.88 | 99.88 | 99.88 | 99.89 | 99.89 | 99.90 | 99.90 |
| 3.30 | 99.90 | 99.91 | 99.91 | 99.91 | 99.92 | 99.92 | 99.92 | 99.92 | 99.93 | 99.93 |
| 3.40 | 99.93 | 99.94 | 99.94 | 99.94 | 99.94 | 99.94 | 99.94 | 99.95 | 99.95 | 99.95 |
| 3.50 | 99.95 | 99.96 | 99.96 | 99.96 | 99.96 | 99.96 | 99.96 | 99.96 | 99.97 | 99.97 |

TABLE 3.3
ONE-TAILED GAUSSIAN INTEGRAL
Giving the percentage probability that a point lies within the given number of $\sigma$ to one side of the mean

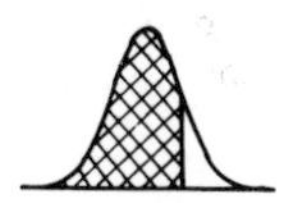

| $\frac{x-\mu}{\sigma}$ | 0.00 | 0.01 | 0.02 | 0.03 | 0.04 | 0.05 | 0.06 | 0.07 | 0.08 | 0.09 |
|---|---|---|---|---|---|---|---|---|---|---|
| 0.00 | 50.00 | 50.40 | 50.80 | 51.20 | 51.60 | 51.99 | 52.39 | 52.79 | 53.19 | 53.59 |
| 0.10 | 53.98 | 54.38 | 54.78 | 55.17 | 55.57 | 55.96 | 56.36 | 56.75 | 57.14 | 57.53 |
| 0.20 | 57.93 | 58.32 | 58.71 | 59.10 | 59.48 | 59.87 | 60.26 | 60.64 | 61.03 | 61.41 |
| 0.30 | 61.79 | 62.17 | 62.55 | 62.93 | 63.31 | 63.68 | 64.06 | 64.43 | 64.80 | 65.17 |
| 0.40 | 65.54 | 65.91 | 66.28 | 66.64 | 67.00 | 67.36 | 67.72 | 68.08 | 68.44 | 68.79 |
| 0.50 | 69.15 | 69.50 | 69.85 | 70.19 | 70.54 | 70.88 | 71.23 | 71.57 | 71.90 | 72.24 |
| 0.60 | 72.57 | 72.91 | 73.24 | 73.57 | 73.89 | 74.22 | 74.54 | 74.86 | 75.17 | 75.49 |
| 0.70 | 75.80 | 76.11 | 76.42 | 76.73 | 77.04 | 77.34 | 77.64 | 77.94 | 78.23 | 78.52 |
| 0.80 | 78.81 | 79.10 | 79.39 | 79.67 | 79.95 | 80.23 | 80.51 | 80.78 | 81.06 | 81.33 |
| 0.90 | 81.59 | 81.86 | 82.12 | 82.38 | 82.64 | 82.89 | 83.15 | 83.40 | 83.65 | 83.89 |
| 1.00 | 84.13 | 84.38 | 84.61 | 84.85 | 85.08 | 85.31 | 85.54 | 85.77 | 85.99 | 86.21 |
| 1.10 | 86.43 | 86.65 | 86.86 | 87.08 | 87.29 | 87.49 | 87.70 | 87.90 | 88.10 | 88.30 |
| 1.20 | 88.49 | 88.69 | 88.88 | 89.07 | 89.25 | 89.44 | 89.62 | 89.80 | 89.97 | 90.15 |
| 1.30 | 90.32 | 90.49 | 90.66 | 90.82 | 90.99 | 91.15 | 91.31 | 91.47 | 91.62 | 91.77 |
| 1.40 | 91.92 | 92.07 | 92.22 | 92.36 | 92.51 | 92.65 | 92.79 | 92.92 | 93.06 | 93.19 |
| 1.50 | 93.32 | 93.45 | 93.57 | 93.70 | 93.82 | 93.94 | 94.06 | 94.18 | 94.29 | 94.41 |
| 1.60 | 94.52 | 94.63 | 94.74 | 94.84 | 94.95 | 95.05 | 95.15 | 95.25 | 95.35 | 95.45 |
| 1.70 | 95.54 | 95.64 | 95.73 | 95.82 | 95.91 | 95.99 | 96.08 | 96.16 | 96.25 | 96.33 |
| 1.80 | 96.41 | 96.49 | 96.56 | 96.64 | 96.71 | 96.78 | 96.86 | 96.93 | 96.99 | 97.06 |
| 1.90 | 97.13 | 97.19 | 97.26 | 97.32 | 97.38 | 97.44 | 97.50 | 97.56 | 97.61 | 97.67 |
| 2.00 | 97.72 | 97.78 | 97.83 | 97.88 | 97.93 | 97.98 | 98.03 | 98.08 | 98.12 | 98.17 |
| 2.10 | 98.21 | 98.26 | 98.30 | 98.34 | 98.38 | 98.42 | 98.46 | 98.50 | 98.54 | 98.57 |
| 2.20 | 98.61 | 98.64 | 98.68 | 98.71 | 98.75 | 98.78 | 98.81 | 98.84 | 98.87 | 98.90 |
| 2.30 | 98.93 | 98.96 | 98.98 | 99.01 | 99.04 | 99.06 | 99.09 | 99.11 | 99.13 | 99.16 |
| 2.40 | 99.18 | 99.20 | 99.22 | 99.25 | 99.27 | 99.29 | 99.31 | 99.32 | 99.34 | 99.36 |
| 2.50 | 99.38 | 99.40 | 99.41 | 99.43 | 99.45 | 99.46 | 99.48 | 99.49 | 99.51 | 99.52 |
| 2.60 | 99.53 | 99.55 | 99.56 | 99.57 | 99.59 | 99.60 | 99.61 | 99.62 | 99.63 | 99.64 |
| 2.70 | 99.65 | 99.66 | 99.67 | 99.68 | 99.69 | 99.70 | 99.71 | 99.72 | 99.73 | 99.74 |
| 2.80 | 99.74 | 99.75 | 99.76 | 99.77 | 99.77 | 99.78 | 99.79 | 99.79 | 99.80 | 99.81 |
| 2.90 | 99.81 | 99.82 | 99.82 | 99.83 | 99.84 | 99.84 | 99.85 | 99.85 | 99.86 | 99.86 |
| 3.00 | 99.87 | 99.87 | 99.87 | 99.88 | 99.88 | 99.89 | 99.89 | 99.89 | 99.90 | 99.90 |
| 3.10 | 99.90 | 99.91 | 99.91 | 99.91 | 99.92 | 99.92 | 99.92 | 99.92 | 99.93 | 99.93 |
| 3.20 | 99.93 | 99.93 | 99.94 | 99.94 | 99.94 | 99.94 | 99.94 | 99.95 | 99.95 | 99.95 |
| 3.30 | 99.95 | 99.95 | 99.95 | 99.96 | 99.96 | 99.96 | 99.96 | 99.96 | 99.96 | 99.97 |
| 3.40 | 99.97 | 99.97 | 99.97 | 99.97 | 99.97 | 99.97 | 99.97 | 99.97 | 99.97 | 99.98 |
| 3.50 | 99.98 | 99.98 | 99.98 | 99.98 | 99.98 | 99.98 | 99.98 | 99.98 | 99.98 | 99.98 |

The $2\sigma$ value is so close to 95% (and vice versa) that in practice the difference can often be ignored. From the $1\sigma$ value you obtain the useful rule of thumb that when a curve is shown going through a set of measured points with error bars, about one third of the error bars should miss the curve. Many people fail to realise this and overestimate their errors in an effort to make the curve go through all the points. It is thus a standard ploy in seminars, etc., when hapless speakers proudly present fitted data, to attack them for having too good a fit.

Sometimes you are interested in the probability of a value straying in one direction only—for example, you may want to be sure that some upper limit is not exceeded, but do not care how far it strays below the mean. For this you need the *one-tailed* probability, as shown in Table 3.3, as opposed to the *two-tailed* probability of Table 3.2.

Should you ever need to know the integrated Gaussian for any other (asymmetric) limits, it can be obtained from these tables by simple arithmetic. Indeed, Tables 3.2 and 3.3 can readily be obtained from each other, but both are given here for convenience of use.

### 3.4.5 Gaussian as Limit of the Poisson and Binomial

From the distributions shown Figure 3.2 it can be seen that for large $\lambda$, the Poisson distribution tends to a Gaussian shape, with $\mu = \lambda$, $\sigma = \sqrt{\lambda}$. In such cases the Gaussian may be used as a very convenient approximation to the Poisson. What is 'large' depends on how good an agreement you require. Some people put the requirement as low as $\lambda = 5$, but 10 is probably safer.

Proof: let $r = \lambda + x$, and use Stirling's approximation:

$$\ln r! \approx r \ln r - r + \ln \sqrt{(2\pi r)}.$$

Then, taking the logarithm of equation 3.11,

$$\ln P(r; \lambda) \approx -\lambda + r \ln \lambda - (r \ln r - r) - \ln \sqrt{2\pi r}$$

$$\approx -\lambda + r\left\{\ln \lambda - \ln\left[\lambda\left(1 + \frac{x}{\lambda}\right)\right]\right\} + (\lambda + x) - \ln \sqrt{2\pi\lambda}.$$

Using the expansion $\ln(1 + z) = z - z^2/2 \ldots$,

$$\ln P(r; \lambda) \approx x - (\lambda + x)\left(\frac{x}{\lambda} - \frac{x^2}{2\lambda^2}\right) - \ln \sqrt{2\pi\lambda}$$

$$\approx -\frac{x^2}{2\lambda} - \ln \sqrt{2\pi\lambda}.$$

Thus, exponentiating,

$$P(x) = \frac{e^{-x^2/2\lambda}}{\sqrt{2\pi\lambda}}.$$

*Example Poisson approximated by Gaussian*
If $\lambda$ is 5.3, then the probability of two events or less is 10.2%, using the Poisson formula. Approximating the histogram of the Poisson by the smooth Gaussian curve, the appropriate value for the Gaussian is halfway between the possible discrete values of 2 and 3, at 2.5 'events'. This is $(5.3-2.5)/\sqrt{5.3}=1.22\sigma$ from the mean, and Table 3.3 gives this one-tailed probability as 11.1%.

Likewise the binomial tends to a Gaussian with $\mu=np$ and $\sigma=\sqrt{np(1-p)}$. (The proof is similar to that for the Poisson.) This happens first for $p\approx 0.5$; large or small values of $p$ require a larger $n$. Indeed, almost everything tends to a Gaussian as the numbers become large—this is due to the *central limit theorem*, discussed in the next chapter.

### ★ 3.4.6 The Many-dimensional Gaussian

Consider a distribution in $n$ variables, denoted by $x_{(1)}, x_{(2)},\ldots,x_{(n)}$—the notation is discussed in section 2.6.3. These can be written compactly as a vector $\mathbf{x}$, likewise the means, $\mu_{(1)}, \mu_{(2)},\ldots,\mu_{(n)}$ can be written $\boldsymbol{\mu}$. The most general form of multi-dimensional Gaussian is an exponential of a quadratic form, which will contain terms in $x_{(i)}^2$, cross terms in $x_{(i)}x_{(j)}$, linear terms, and a constant, but nothing of higher power. This can be written:

$$P(\mathbf{x})\propto \exp\left[-\frac{1}{2}(\tilde{\mathbf{x}}-\tilde{\boldsymbol{\mu}})\mathbf{A}(\mathbf{x}-\boldsymbol{\mu})\right].$$

Even this contains some ambiguity, which can be resolved by insisting that $\mathbf{A}$ be symmetric:

$$A_{ij}=A_{ji}.$$

Henceforth suppose, without loss of generality, that all $\mu_{(i)}$ are zero.

It may be that $\mathbf{A}$ is diagonal, i.e. all the cross terms are zero. In that case $P(\mathbf{x})$ factorises into $n$ independent Gaussians:

$$e^{-(A_{11}x_1^2+A_{22}x_2^2+A_{33}x_3^2+\cdots)/2}=e^{-A_{11}x_1^2/2}\,e^{-A_{22}x_2^2/2}\,e^{-A_{33}x_3^2/2}\cdots$$

and the diagonal elements can be identified as

$$A_{ii}=\frac{1}{\sigma_i^2}.$$

As $\mathbf{A}$ is diagonal this can be written in the form

$$\mathbf{A}=\mathbf{V}^{-1}. \tag{3.22}$$

Now we go on to consider the general case and to show that the above equation is still true. Even if $\mathbf{A}$ is not diagonal, a unitary matrix $\mathbf{U}$ can always be found to diagonalise it; i.e.

$$\mathbf{U}\mathbf{A}\tilde{\mathbf{U}}=\mathbf{A}', \quad \text{where } \mathbf{A}' \text{ is diagonal.}$$

Note: 'unitary' means that the transposed matrix is the same as the inverse: $\mathbf{U}^{-1} = \tilde{\mathbf{U}}$. The significance of this is that if one considers vectors $\mathbf{x}, \mathbf{y}, \ldots$ transformed by $\mathbf{U}$

$$\mathbf{x}' = \mathbf{U}\mathbf{x} \quad \mathbf{y}' = \mathbf{U}\mathbf{y} \quad \text{etc.}$$

then the transposes (denoted by a tilde, ˜) are given by

$$\tilde{\mathbf{x}}' = \tilde{\mathbf{x}}\tilde{\mathbf{U}}$$

so the 'scalar product' of two vectors does not change under transformations by a unitary matrix,

$$\tilde{\mathbf{x}}'\mathbf{y}' = \tilde{\mathbf{x}}\tilde{\mathbf{U}}\mathbf{U}\mathbf{y} = \tilde{\mathbf{x}}\mathbf{U}^{-1}\mathbf{U}\mathbf{y} = \tilde{\mathbf{x}}\mathbf{y}$$

and they thus represent generalised rotations.

It is a basic fact of matrix algebra that for any symmetric matrix a unitary matrix can always be found for which $\mathbf{U}\mathbf{A}\tilde{\mathbf{U}}$ is diagonal.

The exponent $\tilde{\mathbf{x}}\mathbf{A}\mathbf{x}$ can be written

$$\tilde{\mathbf{x}}\tilde{\mathbf{U}}\mathbf{U}\mathbf{A}\tilde{\mathbf{U}}\mathbf{U}\mathbf{x}.$$

This is $\mathbf{x}'\mathbf{A}'\mathbf{x}'$, with $\mathbf{A}'$ diagonal. The variance matrix $\mathbf{V}'$, for the $\mathbf{x}'$, is thus diagonal with elements $(\mathbf{U}\mathbf{A}\tilde{\mathbf{U}})^{-1} = \mathbf{U}\mathbf{A}^{-1}\tilde{\mathbf{U}}$, by equation 3.22.

So we know the variance matrix for the $\mathbf{x}'$, and also that the $\mathbf{x}$ are related to these by $\mathbf{x} = \tilde{\mathbf{U}}\mathbf{x}'$. We now (anticipating a result from the next chapter) invoke the generalised combination of errors formula, equation 4.19, which gives the variance matrix for a set of variables which are a function of another set. (Incidentally, as in this case the relation is linear, the equation is exact and not an approximation.) The derivative matrix $\mathbf{G}$ in 4.19 is just $\mathbf{U}$, so

$$\begin{gathered} \mathbf{V} = \tilde{\mathbf{U}}\mathbf{V}'\mathbf{U} = \tilde{\mathbf{U}}\mathbf{U}\mathbf{A}^{-1}\tilde{\mathbf{U}}\mathbf{U} \\ \mathbf{V} = \mathbf{A}^{-1}. \end{gathered} \tag{3.23}$$

**Result** *The matrix in the exponent of the multidimensional Gaussian is the inverse of the covariance matrix.*

In full, with the normalisation (which can be found from the Jacobian of the $x \to x'$ transformation):

$$P(\mathbf{x}) = \frac{1}{(2\pi)^{n/2}\sqrt{|\mathbf{V}|}} \exp\left[-\tfrac{1}{2}(\tilde{\mathbf{x}} - \tilde{\boldsymbol{\mu}})\mathbf{V}^{-1}(\mathbf{x} - \boldsymbol{\mu})\right].$$

### ★ 3.4.7 The Binormal Distribution

For two dimensions (calling the variables $x$ and $y$ again, rather than $x_{(1)}$

and $x_{(2)}$) the covariance matrix is

$$V = \begin{pmatrix} \sigma_x^2 & \rho\sigma_x\sigma_y \\ \rho\sigma_x\sigma_y & \sigma_y^2 \end{pmatrix}$$

which has the inverse

$$\mathbf{V}^{-1} = \frac{1}{\sigma_x^2\sigma_y^2(1-\rho^2)} \begin{pmatrix} \sigma_y^2 & -\rho\sigma_x\sigma_y \\ -\rho\sigma_x\sigma_y & \sigma_x^2 \end{pmatrix}.$$

The full formula (including normalisation) for the binormal or two-dimensional Gaussian is thus

$$P(x, y) = \frac{1}{2\pi\sigma_x\sigma_y\sqrt{(1-\rho^2)}} \times \exp\left\{-\frac{1}{2(1-\rho^2)}\left[\left(\frac{x-\mu_x}{\sigma_x}\right)^2 + \left(\frac{y-\mu_y}{\sigma_y}\right)^2 - 2\rho\left(\frac{x-\mu_x}{\sigma_x}\right)\left(\frac{y-\mu_y}{\sigma_y}\right)\right]\right\}. \tag{3.24}$$

This can be drawn on paper using contour lines. The contours of equal probability are curves for which the exponent in equation 3.24 is constant, and that is the equation of an ellipse. Manipulation of equation 3.24 shows that the ellipse for which the exponent is $-\frac{1}{2}$ has extreme $x$ and $y$ values at $\mu_x \pm \sigma_x$ and $\mu_y \pm \sigma_y$, i.e. it fits exactly into a rectangular box between these limits.

If you take a slice through the distribution, considering the distribution in $y$, say, for a fixed value of $x$, then, by inspection, equation 3.24 becomes a Gaussian distribution in $y$ whose standard deviation is narrowed to $\sigma_y/\sqrt{1-\rho^2}$ and mean is $\mu_y + \rho(\sigma_y/\sigma_x)(x-\mu_x)$.

In two dimensions the unitary matrix $\mathbf{U}$ that diagonalises the exponent is the familiar rotation matrix

$$\begin{pmatrix} \cos\theta & -\sin\theta \\ \sin\theta & \cos\theta \end{pmatrix}$$

which rotates the $(x, y)$ axes by some angle $\theta$ such that the major and minor axes of these ellipses coincide with the new axes: call them $u$ and $v$. The three parameters of the binormal can thus be written $\sigma_x$, $\sigma_y$, $\rho$, as previously, or as $\sigma_u$, $\sigma_v$, $\theta$, where $u$ and $v$ are uncorrelated and with standard deviations $\sigma_u$ and $\sigma_v$, and the $x$, $y$ system is given by rotating the $u$, $v$ system through an angle $\theta$.

A little algebra gives the relations between the two parameter sets:

$$\tan 2\theta = \frac{2\rho\sigma_x\sigma_y}{\sigma_x^2 - \sigma_y^2}$$

$$\sigma_u^2 = \frac{\cos^2\theta\sigma_x^2 - \sin^2\theta\sigma_y^2}{\cos^2\theta - \sin^2\theta} \qquad \sigma_v^2 = \frac{\cos^2\theta\sigma_y^2 - \sin^2\theta\sigma_x^2}{\cos^2\theta - \sin^2\theta}$$

$$\sigma_x^2 = \cos^2\theta\sigma_u^2 + \sin^2\theta\sigma_v^2 \quad \sigma_y^2 = \cos^2\theta\sigma_v^2 + \sin^2\theta\sigma_u^2$$

$$\rho = \sin\theta\cos\theta\frac{\sigma_u^2 - \sigma_v^2}{\sigma_x\sigma_y}.$$

Figure 3.4 shows the lines of constant probability for a two-dimensional distribution, where $x$ and $y$ are positively correlated. The ellipses of constant probability (at 90, 80,...,10% of the peak value) are shown. The parameters for this figure are

$$\sigma_u = 1.0 \quad \sigma_v = 0.5 \quad \theta = 45°$$

or, equivalently,

$$\sigma_x = \sqrt{\frac{5}{8}} \quad \sigma_y = \sqrt{\frac{5}{8}} \quad \rho = \frac{3}{5}.$$

## ★3.5 OTHER DISTRIBUTIONS

The Gaussian, Poisson, and binomial distributions are, in that order, far and away the most common and useful. However, they are not the only ones, and some others are described here; in addition the *$\chi^2$ distribution*, *Student's t distribution*, and *Fisher's F distribution* will be discussed in later chapters.

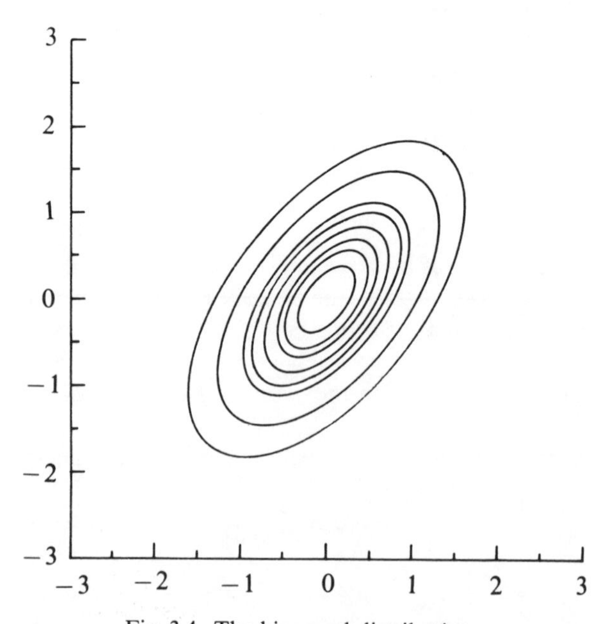


Fig. 3.4. The binormal distribution.

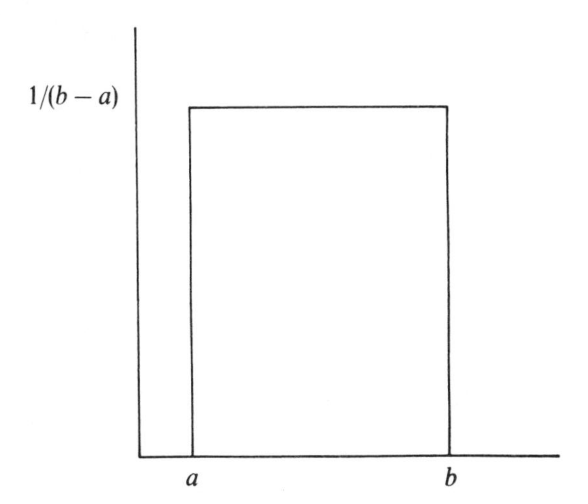


Fig. 3.5. The uniform distribution.

### ★ 3.5.1 The Uniform Distribution

Also known as the rectangular or top hat distribution, the uniform distribution (Figure 3.5) describes a probability which is constant over a certain range and zero outside it. If the range limits are $a$ and $b$ then

$$P(x) = \begin{cases} \dfrac{1}{b-a} & \text{for} \quad a \leqslant x \leqslant b \\ 0 & \text{elsewhere.} \end{cases}$$

The mean is obviously $(a+b)/2$. On doing the integral to obtain $\langle x^2 \rangle$ and thus the variance one gets

$$V(x) = \frac{(b-a)^2}{12} \tag{3.25}$$

i.e. the standard deviation for a uniform distribution is the width divided by $\sqrt{12}$.

### ★ 3.5.2 The Weibull Distribution

$$P(x; \alpha, \beta) = \alpha\beta(\alpha x)^{\beta-1} e^{-(\alpha x)^{\beta}}.$$

Originally invented to describe failure rates in ageing lightbulbs, the Weibull distribution (Figure 3.6) is useful for parametrising functions which rise as $x$ increases from 0 and then fall again. $\alpha$ is just a scale factor. $\beta$ expresses the sharpness of the peak. $\beta = 1$ gives the exponential function.

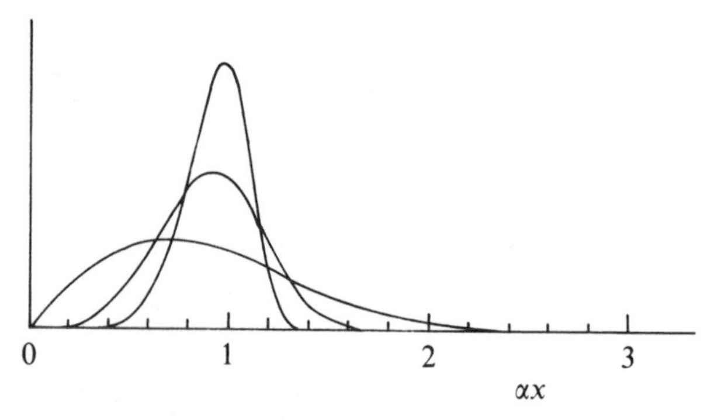


Fig. 3.6. Some Weibull functions. The successively sharper peaks are for $\beta = 2.0$, 4.0, and 7.0.

### ★ 3.5.3 The Breit–Wigner or Cauchy Distribution

$$F(m; M, \Gamma) = \frac{1}{2\pi} \frac{\Gamma}{(m - M)^2 + (\Gamma/2)^2}$$

$$F(z) = \frac{1}{\pi} \frac{1}{1 + z^2}.$$

The Breit–Wigner function, used by nuclear physicists to give the distribution of particles of mass $m$ due to a resonance of mass $M$ and width $\Gamma$, reduces to the Cauchy function $F(z)$ (Figure 3.7) by a change of origin and scale. Its chief feature is its unlovable mathematical behaviour. It does not have a variance as the integral $\int z^2 F(z)\,dz$ diverges.

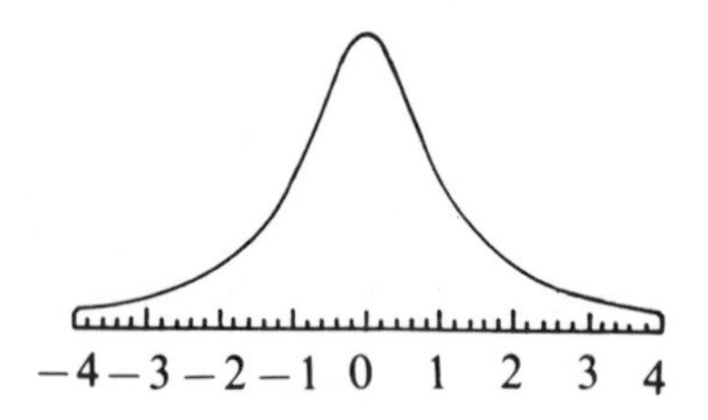


Fig. 3.7. The Cauchy function.

## 3.6 PROBLEMS

*3.1*

A defence system is 99.5% efficient in intercepting ballistic missiles. What is the probability that it will intercept all of 100 missiles launched against it? How many missiles must an aggressor launch to have a better than evens chance of one or more penetrating the defences?

*3.2*

In the previous question, how many missiles would be needed to ensure a better than evens chance of more than two missiles evading the defences?

*3.3*

During a meteor shower, meteors fall at the rate 15.7 per hour. What is the probability of observing less than 5 in a given period of 30 minutes?

*3.4*

Repeat the previous problem, using the Gaussian approximation to the Poisson.

*3.5*

A student is trying to hitch a lift. Cars pass at random intervals, at an average rate of 1 per minute. The probability of a car giving a lift is 1%. What is the probability that the student will still be waiting:

(a) after 60 cars have passed?
(b) after 1 hour?

*3.6*

For a Gaussian distribution, using Tables 3.2 and 3.3:

(a) What is the probability of a value lying more than $1.23\sigma$ from the mean?
(b) What is the probability of a value lying more than $2.43\sigma$ above the mean?
(c) What is the probability of a value lying less than $1.09\sigma$ below the mean?
(d) What is the probability of a value lying above a point $0.45\sigma$ below the mean?
(e) What is the probability that a value lies more than $0.5\sigma$ but less than $1.5\sigma$ from the mean?
(f) What is the probability that a value lies above $1.2\sigma$ on the low side of the mean, and below $2.1\sigma$ on the high side?
(g) Within how many standard deviations does the probability of a value occurring equal 50%?
(h) How many standard deviations correspond to a one-tailed probability of 99%?

★ *3.7*

Show that the skew and curtosis of a Gaussian are zero.

*The errors of a wise man make your rule*
*Rather than the perfections of a fool.*

—*William Blake*

# CHAPTER 4

# Errors

Doing experiments involve making measurements, which are then analysed to produce results. These measurements, whether made by you or me or some eminent Nobel-prizewinning expert, are never perfectly exact, but have some *resolution* or *error*. This chapter deals with the handling of these errors, and how the errors on the measurements combine and propagate through to errors on the results.

When values are quoted with an error, this error is a Gaussian standard deviation $\sigma$. If you say the length of a piece of string is $12.3 \pm 0.1$ cm, you mean that you have measured it with a ruler (or equivalent device) which gives answers that differ from the true value by within 0.1 cm 68% of the time, 0.2 cm 95% of the time, and 0.3 cm 99.7% of the time† (the numbers come from section 3.4.4). This is not just an arbitrary choice. Errors on measurements and results are generally well described by the Gaussian distribution, which is of course why it is also known as the normal distribution.

†Engineers use a different convention. They quote tolerances, and for them $12.3 \pm 0.1$ cm means a guarantee that it lies between 12.2 and 12.4 cm. This can give rise to some amusing communication problems.

## 4.1 WHY ERRORS ARE GAUSSIAN

Measurements acquire errors from many different sources. If you measure the length of a rod using a ruler, all sorts of inexactitudes creep in: optical parallax, the ruler's calibration, rounding errors, your hand shaking, and so on. Reading a meter with a moving pointer has similar problems. Digital meters and electronic readout avoid the effects of shaky hands and bleary eyes, but at the expense of others, hidden in the depths of the electronics. The imperfections in the measurements we make are not due to one cause, but to many.

Now, there is a powerful and surprising result about the behaviour of a variable which is the sum of several others. It is called the *central limit theorem* or *CLT* for short.

**The central limit theorem** *If you take the sum $X$ of $N$ independent variables, $x_i$, where $i = 1, 2, 3, \ldots, N$, each taken from a distribution of mean $\mu_i$ and variance $V_i$ or $\sigma_i^2$, the distribution for $X$*

(a) *has an expectation value*

$$\langle X \rangle = \sum \mu_i \tag{4.1}$$

(b) *has variance*

$$V(X) = \sum V_i = \sum \sigma_i^2 \tag{4.2}$$

(c) *becomes Gaussian as $N \to \infty$.* (4.3)

This is why the Gaussian is so important. A quantity produced by the cumulative effect of many independent variables will be, at least approximately, Gaussian, no matter what the distributions of the original variables may have been. Measurement errors behave accordingly, as do many other observed quantities. For example, human heights are well described by a Gaussian distribution, as are the lengths of the arm, forefinger, and other anatomical measurements, as these are due to the combined effects of many genetic and environmental factors.[†]

Despite this, there is an important piece of small print at the bottom of the contract. The CLT works better in the centre of the distribution than far away from it. Distributions can be indistinguishable from a Gaussian within one or two $\sigma$ of the peak, but deviate in regions farther away. These, with fine disregard for anatomy, are called the 'tails' by some and the 'wings' by others. Whatever you call them, they are dangerous.

[†]Weight, on the other hand, is not Gaussian but has a positive skew, as the single factor of how much you eat dominates the others.

### 4.1.1 The Central Limit Theorem

The proof of equation 4.1 is very simple, as it merely states that the expectation value of a sum is the sum of the expectation values

$$
\begin{aligned}
X &= \sum_i x_i \\
\langle X \rangle &= \left\langle \sum_i x_i \right\rangle \\
&= \sum_i \langle x_i \rangle \\
&= \sum_i \mu_i .
\end{aligned}
$$

To see that equation 4.2 is true, write out the variance and insert equation 4.1.

$$
\begin{aligned}
V(X) &= \langle (X - \langle X \rangle)^2 \rangle \\
&= \left\langle \left( \sum_i x_i - \sum_i \mu_i \right)^2 \right\rangle \\
&= \left\langle \left[ \sum_i (x_i - \mu_i) \right]^2 \right\rangle \\
&= \left\langle \sum_i (x_i - \mu_i)^2 \right\rangle + \left\langle \sum_i \sum_{j \neq i} (x_i - \mu_i)(x_j - \mu_j) \right\rangle \\
&= \sum_i \langle (x_i - \mu_i)^2 \rangle + \sum_i \sum_{j \neq i} \langle (x_i - \mu_i)(x_j - \mu_j) \rangle .
\end{aligned}
$$

The elements of the second sum are the covariances (see section 2.6) between pairs of different measurements, but as these are independent (this is specified in the statement of the theorem) each term is zero. This leaves only the first sum, which is just $\sum V_i$, the desired result.

Notice in passing that if the variables are not independent then equation 4.1 is still valid, but equation 4.2 is not.

Proof of the third and most fundamental part, equation 4.3, is complicated, and has been relegated to Appendix 2. However, it can be demonstrated to work, even under apparently unpromising conditions. Figure 4.1 shows an example.

A. This shows a histogram of 5000 numbers taken at random from a uniform distribution between 0 and 1. It has mean 1/2 and variance 1/12, and is essentially flat—certainly not at all Gaussian.
B. Here are another 5000 numbers, each the sum of a pair of random numbers of the type shown in A, i.e. $X = x_1 + x_2$. The distribution is triangular, peaking at 1.0, and falling linearly to zero at 0 and 2.

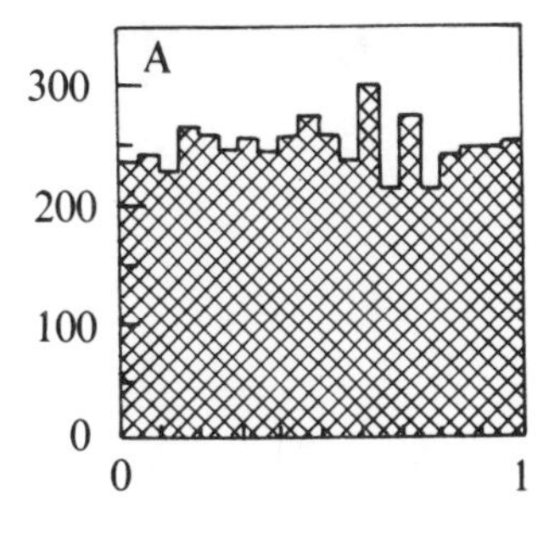


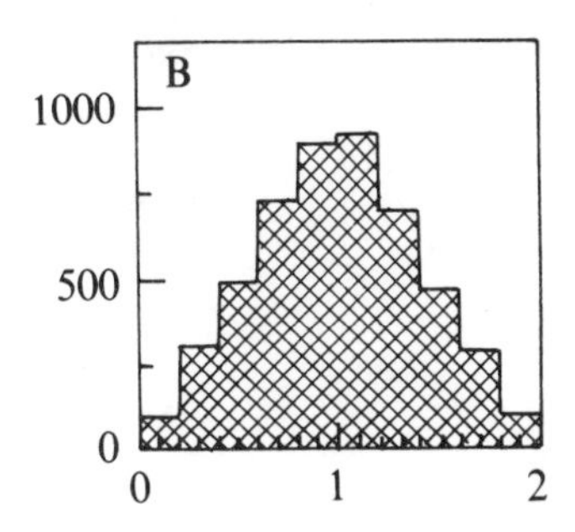


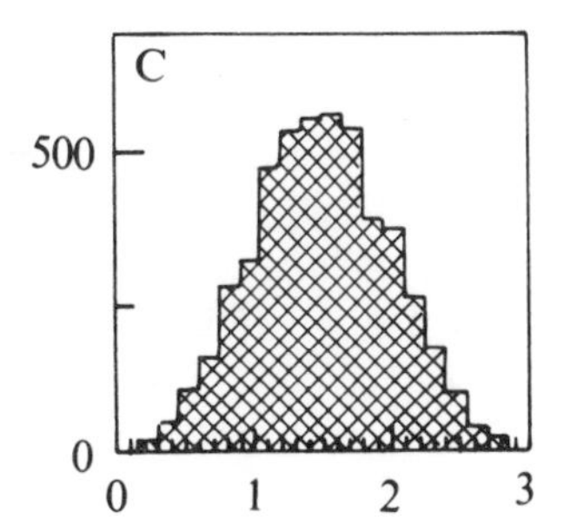


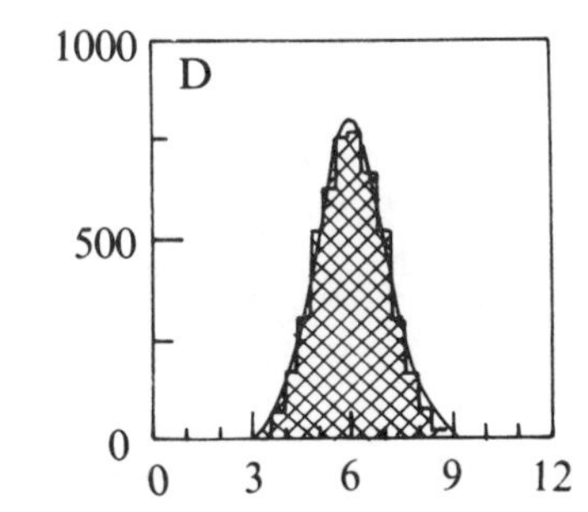


Fig. 4.1. The CLT at work.

C. If three uniformly distributed random numbers are added together, the peak is at 1.5, and the form becomes curved.
D. When the number of uniformly distributed random numbers increases to 12 at a time, the histogram appears to the eye just like a true Gaussian. The Gaussian of mean 6.0 and $\sigma^2 = 12 \times \frac{1}{12} = 1.0$ is also shown, and the agreement is seen to be very good indeed. (Incidently, this provides a convenient but inefficient way of generating random numbers according to a Gaussian distribution.)

## 4.2 WORKING WITH ERRORS

Having sorted out what we mean by errors, we can now deal with some basic ideas that arise working with them.

### 4.2.1 Repeated Measurements

Suppose the same quantity is measured many times. Then the CLT can be used in a simple form as all the $\mu_i$ have the same value—call it $\mu$—and all $\sigma_i$ have the same value $\sigma$. Equation 4.1 reads

$$\langle X \rangle = \sum \mu = N\mu$$

and in terms of the average $\bar{x} = X/N$

$$\langle \bar{x} \rangle = \mu \tag{4.4}$$

and provided the measurements are independent the variance of $\bar{x}$ is just the variance of $X$, divided by $N^2$ (squared as $V$ is a squared quantity):

$$V(\bar{x}) = \frac{1}{N^2} \sum V_i$$

$$= \frac{\sigma^2}{N}. \tag{4.5}$$

> 'What's this thing $\langle \bar{x} \rangle$?' you ask. 'What do you mean by the average average $x$?' Let us spell it out. You take $N$ measurements $x_1, x_2, \ldots, x_N$ and average them: this average is called $\bar{x}$. This result is subject to statistical fluctuations, but on average its value will be $\mu$—sometimes more, sometimes less, but on average it will be $\mu$, i.e. $\langle \bar{x} \rangle = \mu$. The difference between your actual measured $\bar{x}$ and the 'true' value $\mu$ is described by some distribution which has variance $V(\bar{x}) = \sigma^2/N$.

Thus (and this is the payoff line), the standard deviation of this average falls like $1/\sqrt{N}$. This is summarised in the well-known statistical rule 'averaging is good for you'. It is also an example of the law of diminishing returns: to get twice as good a resolution you need four times as many measurements. If you can take $N$ independent measurements of something,

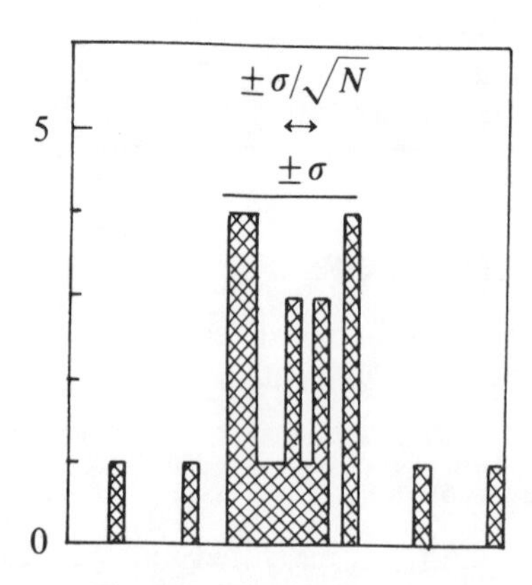


Fig. 4.2. Standard errors.

them their average has an expectation value which is exactly the desired quantity, and as the variance fall like $1/N$, the resolution or 'error' on your average is $\sigma/\sqrt{N}$, which is smaller by a factor $1/\sqrt{N}$ than the error on a single measurement. (It is worth noticing that this holds for any distribution, not just the Gaussian.) $\sigma/\sqrt{N}$ is called the *standard error on the mean*, as it describes how well you know the mean of a distribution, which is often a very important quantity. Notice how it depends on the $\sigma$ of the distribution and also on the number of measurements used.

Figure 4.2 shows a histogram, obtained from randomly sampling a Gaussian distribution. It contains 25 measurements, and $\sigma$ is shown, as is the standard error on the mean, i.e. the resolution with which you can establish the value of the true mean from these 25 measurements, which is smaller than $\sigma$ by a factor of 5.

*Example Photon energy measurements*
The energy resolution of a $\gamma$-ray detector used to investigate a decaying nuclear isotope is 50 keV. If only one such decays is observed, its energy is known to 50 keV. If 100 are collected, this improves to 5 keV. To reach 1 keV you would need to observe 2500 decays.

*Example Weights of eggs*
The weights of eggs produced by a farmer's hens have a standard deviation of 10 g. He feeds a group of hens an expensive vitamin supplement, which will pay back its costs if it increases the weight of the eggs by 2 g. He measures 25 eggs from vitamin-fed hens and their average has increased by 3 g. Does this prove anything useful?

No. The standard error on the mean is $10/\sqrt{25} = 2$ g, so the increase is only $1.5\sigma$ and is not really significant.

Notice the subtle difference in this example, in that the variation in the egg weights comes from the spread in the objects themselves, not from the measurement process; it is assumed that the eggs can be weighed with complete accuracy, or at any rate an accuracy much better than 10 g, which is probably true.

### 4.2.2 Averaging Weighted Measurements

Suppose you have a set of measurements $\{x_i\}$ of some quantity $\mu$ and that these measurements have different error $\sigma_i$. To combine the values you obviously want to form an average in such a way that the better measurements (i.e. those with small $\sigma$) are given more weight than the poorer, large $\sigma$, measurements.

To give a specific example, suppose that a voltage has been measured as 3.11 volts by a meter whose resolution is known to be 0.02 volts, and 3.13 volts by another, better, meter whose resolution is known to be 0.01 volts. How can these be usefully combined to give a single result?

Well, if you had taken four measurements with the poorer, 0.02 volt resolution, meter, and averaged them, then the average would have had a precision of $0.02/\sqrt{4} = 0.01$ volts. Four poor measurements are equivalent to one measurement twice as good, i.e. with half the error. You can run this argument backwards: one of the good measurements is equivalent to four of the inferior ones, and should thus be given four times the weight. This gives a result of

$$\bar{V} = \frac{1}{5} \times 3.11 + \frac{4}{5} \times 3.13 = 3.126 \text{ volts.}$$

Generalising, the weight you give to a result when averaging should be proportional to the inverse square of the resolution. When dealing with measurements of the same quantity, each with an error $\sigma_i$, the correct average to form is

$$\bar{x} = \frac{\sum x_i/\sigma_i^2}{\sum 1/\sigma_i^2}. \tag{4.6}$$

What about the resolution of this average? In this case it is obviously $0.02/\sqrt{5} = 0.009$, which generalises (you can justify this from the combination of errors formula—see next section) to a variance of

$$V(\bar{x}) = \frac{1}{\sum 1/\sigma_i^2}. \tag{4.7}$$

*So unexpected was the hole that for several years computers analysing ozone data had systematically thrown out the readings that should have pointed to its growth.*

—*New Scientist, 31 March 1988*

### 4.2.3 A Note of Caution

Averaging results, whether weighted or not, needs to be done with due caution and commensense. Even though a measurement has a small quoted error it can still be, not to put too fine a point on it, wrong. If two results are in blatant and obvious disagreement, any average is meaningless and there is no point in performing it. Other cases may be less outrageous, and it may not be clear whether the difference is due to incompatibility or just unlucky chance.

Certainly at some stage you are sure to have a result which disagrees with the rest by many (i.e. three or more) standard deviations. What is the correct way to deal with it?

The first thing to do is go back as far as you can and check the readings. You are very likely to find a misplaced decimal point, or a pair of numbers transposed in the notebook. If you can easily retake the measurement then this should be done—and the moral is to plot your points as you go, so that you can catch these rogues at an early stage, before their origins get lost in the mists of history.

If you cannot find an obvious mistake, then you probably have no choice but to throw the point away. However you should always do so with reluctance. If you have several such points, and/or if there are more points than you would expect with large ($> 2\sigma$) deviations, then you should be extremely suspicious, as there is probably some effect at work that you do not understand, and you should understand. It is usually a trivial matter, but it could be something new and fundamental. Distrust all algorithms that advise the automatic rejection of points outside certain limits as they can rapidly get out of hand; points should only be condemned after giving them a fair hearing.

## 4.3 COMBINATION OF ERRORS

The measurements you make are analysed and proceed to give results. The inescapable errors on your measurements thus give rise to inescapable errors on your results. This section deals with the way that errors combine and propagate to give errors on the final answer.

### 4.3.1 One Variable

Suppose that $f$ is a simple linear function of $x$:

$$f = ax + b$$

where $a$ and $b$ are exact constants and $x$ has some distribution with variance $V(x)$ or, equivalently, error $\sigma_x$. $x$ represents a measurement, or perhaps an intermediate result in the analysis, and $f$ could be the final result or another intermediate step.

The variance of $f$ is given by

$$\begin{aligned}
V(f) &= \langle f^2 \rangle - \langle f \rangle^2 \\
&= \langle (ax+b)^2 \rangle - \langle ax+b \rangle^2 \\
&= a^2\langle x^2 \rangle + 2ab\langle x \rangle + b^2 - a^2\langle x \rangle^2 - 2ab\langle x \rangle - b^2 \\
&= a^2(\langle x^2 \rangle - \langle x \rangle^2) \\
&= a^2 V(x). \qquad (4.8a)
\end{aligned}$$

This can also be written in terms of the standard deviations as

$$\sigma_f = |a| \sigma_x. \tag{4.8b}$$

This makes good sense. $b$ is a constant, and adding it to a variable does nothing to the spread. $a$ just multiplies the whole distribution by a factor, and increases the width accordingly. It also keeps the dimensions straight if necessary.

Now consider the more useful case where $f$ is some general function of $x$. For small differences we can expand in a Taylor series about some point $x_0$:

$$f(x) \approx f(x_0) + (x - x_0)\left(\frac{\mathrm{d}f}{\mathrm{d}x}\right)\bigg|_{x=x_0}$$

and applying equations 4.8 give

$$V(f) \approx \left(\frac{\mathrm{d}f}{\mathrm{d}x}\right)^2 V(x) \tag{4.9}$$

$$\sigma_f \approx \left|\frac{\mathrm{d}f}{\mathrm{d}x}\right| \sigma_x. \tag{4.10}$$

The approximation is valid for 'small' errors, and 'small' here means that the first differential does not change much over a few $\sigma$. The differentials should be evaluated at the true value of $x$. If this is unknown then the measured value is used, but the difference between the two is insignificant for just this reason.

*Example Distance and time*
If the speed of a projectile is given as $200 \pm 10$ m/sec, the distance travelled in 6 seconds is $1200 \pm 60$ m.

*Example Trigonometry*
If an angle $\theta$ is known with an error of 0.01 radians, then $\sin(\theta)$ is known with an error of $0.01\,|\cos(\theta)|$.

### 4.3.2 A Function of Two or More Variables

Now suppose that $f$ is a function of two variables, $x$ and $y$. As in the previous case, consider first a linear relation

$$f = ax + by + c$$

where $a, b, c$ are constants. Expanding as before gives

$$\begin{aligned} V(f) &= a^2(\langle x^2 \rangle - \langle x \rangle^2) + b^2(\langle y^2 \rangle - \langle y \rangle^2) \\ &\quad + 2ab(\langle xy \rangle - \langle x \rangle \langle y \rangle) \\ &= a^2 V(x) + b^2 V(y) + 2ab \operatorname{cov}(x, y). \end{aligned}$$

Applying this to the more general $f(x, y)$, again for 'small' errors, gives

$$V(f) = \left(\frac{\mathrm{d}f}{\mathrm{d}x}\right)^2 V(x) + \left(\frac{\mathrm{d}f}{\mathrm{d}y}\right)^2 V(y) + 2\left(\frac{\mathrm{d}f}{\mathrm{d}x}\right)\left(\frac{\mathrm{d}f}{\mathrm{d}y}\right)\cos(x, y) \tag{4.11}$$

$$\sigma_f^2 = \left(\frac{\mathrm{d}f}{\mathrm{d}x}\right)^2 \sigma_x^2 + \left(\frac{\mathrm{d}f}{\mathrm{d}y}\right)^2 \sigma_y^2 + 2\left(\frac{\mathrm{d}f}{\mathrm{d}x}\right)\left(\frac{\mathrm{d}f}{\mathrm{d}y}\right)\rho\sigma_x\sigma_y \tag{4.12}$$

where the differentials are evaluated as before at the true or measured values of $(x, y)$.

If, and only if, $x$ and $y$ are uncorrelated, then the covariance term vanishes and 4.11 reduces to the well-known formula for the combination (or propagation) or errors.

**The law of combination of errors** *For x, y independent*

$$V(f) = \left(\frac{\mathrm{d}f}{\mathrm{d}x}\right)^2 V(x) + \left(\frac{\mathrm{d}f}{\mathrm{d}y}\right)^2 V(y) \tag{4.13}$$

$$\sigma_f^2 = \left(\frac{\mathrm{d}f}{\mathrm{d}x}\right)^2 \sigma_x^2 + \left(\frac{\mathrm{d}f}{\mathrm{d}y}\right)^2 \sigma_y^2. \tag{4.14}$$

This says that the error on $x$ and $y$, multiplied by suitable scaling factors, are added in quadrature. Adding two positive numbers in quadrature gives a smaller result than the usual arithmetic addition. Intuitively you can see that this is reasonable: errors of overestimation in $x$ have a fair chance of being compensated by errors of underestimation in $y$.

The extension to more than two variables is straightforward; all you do (provided they are all independent, of course) is add the errors in quadrature. Thus for a function of $x, y$, and $z$,

$$\sigma_f^2 = \left(\frac{\partial f}{\partial x}\right)^2 \sigma_x^2 + \left(\frac{\partial f}{\partial y}\right)^2 \sigma_y^2 + \left(\frac{\partial f}{\partial z}\right)^2 \sigma_z^2.$$

*Example Distance and time (2)*
If a body is moving with speed $200 \pm 10$ m/sec, and accelerating at $12 \pm 2$ m/sec$^2$, then in 6 seconds it travels (using $s = ut + \frac{1}{2}at^2$) a distance of 1416 m. There is an error on this figure of $\pm 60$ m due to the uncertainty in velocity, and $\pm 36$ m due to the uncertainty in acceleration: these add in quadrature to give a combined uncertainty of $\pm 70$ m.

*Example More trigonometry*
If $y = A\sin\theta + B\cos\theta$, then the error on $y$ is given by

$$\sigma_y^2 = \sin^2\theta\sigma_A^2 + \cos^2\theta\sigma_B^2 + (A\cos\theta - B\sin\theta)^2\sigma_\theta^2$$

provided $A, B$, and $\theta$ are independent.

### 4.3.3 Percentage Errors

Several results can be expressed neatly using percentage or fractional errors. Applying the law of combination of errors, equation 4.13, to the product of two independent variables

$$f = xy$$

gives

$$V(f) = y^2 V(x) + x^2 V(y)$$

which can be nicely rewritten as

$$\left(\frac{\sigma_f}{f}\right)^2 = \left(\frac{\sigma_x}{x}\right)^2 + \left(\frac{\sigma_y}{y}\right)^2. \tag{4.15}$$

This also work for quotients. If

$$f = \frac{x}{y}$$

then applying equation 4.13 also leads to equation 4.15. For products and for quotients, the percentage or fraction errors add in quadrature. If you know $x$ to 3% and $y$ to 4%, then you know $xy$ and $x/y$ and $y/x$ to 5%.

Percentage errors are also useful for reciprocals. Equation 4.10 shows that the percentage error on a quantity and its reciprocal are the same:

$$\frac{\sigma_{1/x}}{1/x} = \frac{\sigma_x}{x}. \tag{4.16}$$

Furthermore, the error on the logarithm of $x$ is just the fractional error

$$\sigma_{\ln x} = \frac{\sigma_x}{x}. \tag{4.17}$$

*Example Ohm's law*
The current produced by a voltage $V \pm \sigma_V$ through a resistance $R \pm \sigma_R$ is $I = V/R$ with uncertainty given by

$$\left(\frac{\sigma_I}{I}\right)^2 = \left(\frac{\sigma_V}{V}\right)^2 + \left(\frac{\sigma_R}{R}\right)^2.$$

### ★ 4.3.4 Several Functions of Several Variables

The covariance and correlation between measurements in a data sample are discussed in section 2.6.3. We consider covariance and correlation of variables described by probability distributions. If there are $n$ variables they are denoted $x_{(1)}, x_{(2)}, \ldots, x_{(n)}$.† The covariance between two variables in a

†The parentheses round the subscripts are used to distinguish between the $n$ elements of a single measurement, as opposed to the $N$ measurements that comprise the sample.

sample was defined in section 2.6.3 as

$$\operatorname{cov}(x_{(i)}, x_{(j)}) = \overline{x_{(i)}x_{(j)}} - \overline{x_{(i)}}\,\overline{x_{(j)}}.$$

The covariance between two variables in a joint distribution function $P(x_{(1)}, x_{(2)}, x_{(3)}, \ldots, x_{(n)})$ is exactly equivalent:

$$\operatorname{cov}(x_{(i)}, x_{(j)}) = \langle (x_{(i)} - \mu_i)(x_{(j)} - \mu_j) \rangle$$
$$= \langle x_{(i)}x_{(j)} \rangle - \mu_i\mu_j$$

where $\mu_i$ is the same as $\langle x_{(i)} \rangle$. These are the elements of the *covariance matrix*, also known as the *error matrix*, $\mathbf{V}$, where

$$V_{ij} = \operatorname{cov}(x_{(i)}, x_{(j)}).$$

The diagonal elements of this matrix are just the variances

$$V_{ii} = \operatorname{cov}(x_{(i)}, x_{(i)}) = V(x_{(i)}) = \sigma_i^2.$$

The *correlation matrix* is the dimensionless equivalent

$$\rho_{ij} = \frac{\operatorname{cov}(x_{(i)}, x_{(j)})}{\sigma_i \sigma_j}.$$

Its elements must lie between $-1$ and $+1$, and give the extent to which two variables are correlated.

Now suppose there are $m$ different function $f_1, f_2, f_3, \ldots, f_m$, of $n$ different variables $x_{(1)}, x_{(2)}, x_{(3)}, \ldots, x_{(n)}$. The $x_{(i)}$ have associated errors/variances, therefore so do the $f_k$. The $f_k$ will be correlated with one another (even if the $x_{(i)}$ are not) because the different $f_k$ share the same $x_{(i)}$. The variances on the $f_k$ are given by

$$V(f_i) = \langle f_i^2 \rangle - \langle f_i \rangle^2.$$

Expanding the $f_i$ in a Taylor series produces

$$f_i \approx f_i(\mu_1, \mu_2, \ldots) + \left(\frac{\partial f_i}{\partial x_1}\right)(x_1 - \mu_1) + \left(\frac{\partial f_i}{\partial x_2}\right)(x_2 - \mu_2) + \cdots$$

and inserting this in the formula for the variance, just as was done in the derivations in sections 4.3.1 and 4.3.2, gives

$$V(f_i) = \left(\frac{\partial f_i}{\partial x_1}\right)^2 \langle (x_1 - \mu_1)^2 \rangle + \cdots + 2\left(\frac{\partial f_i}{\partial x_1}\right)\left(\frac{\partial f_i}{\partial x_2}\right)\langle (x_1 - \mu_1)(x_2 - \mu_2) \rangle \cdots$$
$$= \sum_j \left(\frac{\partial f_i}{\partial x_j}\right)^2 V(x_j) + \sum_j \sum_{k \neq j} \left(\frac{\partial f_i}{\partial x_j}\right)\left(\frac{\partial f_i}{\partial x_k}\right) \operatorname{cov}(x_j, x_k)$$

which is the standard combination of errors formula, equation 4.11, slightly generalized. The covariance of $f_k$ and $f_l$ is found in the same way as

$$\langle f_k f_l \rangle - \langle f_k \rangle \langle f_l \rangle \approx \langle (x_1 - \mu_1)(x_1 - \mu_1) \rangle \left( \frac{\partial f_k}{\partial x_1} \right) \left( \frac{\partial f_l}{\partial x_1} \right) + \cdots$$

$$+ \langle (x_1 - \mu_1)(x_2 - \mu_2) \rangle \left( \frac{\partial f_k}{\partial x_1} \right) \left( \frac{\partial f_l}{\partial x_2} \right) + \cdots$$

which is expressible in summation notation as

$$\operatorname{cov}(f_k, f_l) = \sum_i \sum_j \left( \frac{\partial f_k}{\partial x_i} \right) \left( \frac{\partial f_l}{\partial x_j} \right) \operatorname{cov}(x_i, x_j) \qquad (4.18)$$

which includes the previous formula for the variance as $\operatorname{cov}(f_i, f_i) \equiv V(f_i)$. It can be written completely using matrices – if

$$G_{ki} = \left( \frac{\partial f_k}{\partial x_i} \right)$$

and $\mathbf{V}_\mathbf{x}$ and $\mathbf{V}_\mathbf{f}$ are the error matrices for $\mathbf{x}$ and $\mathbf{f}$ respectively, then equation 4.18 can be written as

$$\mathbf{V}_\mathbf{f} = \mathbf{G} \mathbf{V}_\mathbf{x} \tilde{\mathbf{G}} \qquad (4.19)$$

a neat little formula which contains all there is to know about error propagation.

$\mathbf{V}_\mathbf{x}$ and $\mathbf{V}_\mathbf{f}$ are symmetric and square, of size $n \times n$ and $m \times m$. $\mathbf{G}$ is rectangular, $m \times n$.

*Example Three variables*

If there is just one function $f$, of three variables, $x, y, z$, which are uncorrelated and measured with errors $\sigma_x, \sigma_y, \sigma_z$, then

$$G = \left( \frac{\partial f}{\partial x}, \frac{\partial f}{\partial y}, \frac{\partial f}{\partial z} \right)$$

$$V_f = \left( \frac{\partial f}{\partial x}, \frac{\partial f}{\partial y}, \frac{\partial f}{\partial z} \right) \begin{pmatrix} \sigma_x^2 & 0 & 0 \\ 0 & \sigma_y^2 & 0 \\ 0 & 0 & \sigma_z^2 \end{pmatrix} \begin{pmatrix} \frac{\partial f}{\partial x} \\ \frac{\partial f}{\partial y} \\ \frac{\partial f}{\partial x} \end{pmatrix}$$

$$= \left( \frac{\partial f}{\partial x} \right)^2 \sigma_x^2 + \left( \frac{\partial f}{\partial y} \right)^2 \sigma_y^2 + \left( \frac{\partial f}{\partial z} \right)^2 \sigma_z^2.$$

*Example Three-dimensional geometry*

Tracking chambers measure the position of a particle in cylindrical polar coordinates, $(r, \phi, z)$. $r$ is known very precisely and the error can be taken as zero. $\phi$ and $z$ are measured with uncorrelated errors $\sigma_\phi$ and $\sigma_z$.

What are the errors on the cartesian coordinates $x, y, z$?

Well, $x = r\cos\phi$ and $y = r\sin\phi$ (and $z = z$) so

$$G = \begin{pmatrix} \cos\phi & -r\sin\phi & 0 \\ \sin\phi & r\cos\phi & 0 \\ 0 & 0 & 1 \end{pmatrix}$$

$$\begin{aligned} V_{xyz} &= \begin{pmatrix} \cos\phi & -r\sin\phi & 0 \\ \sin\phi & r\cos\phi & 0 \\ 0 & 0 & 1 \end{pmatrix} \begin{pmatrix} 0 & 0 & 0 \\ 0 & \sigma_\phi^2 & 0 \\ 0 & 0 & \sigma_z^2 \end{pmatrix} \begin{pmatrix} \cos\phi & \sin\phi & 0 \\ -r\sin\phi & r\cos\phi & 0 \\ 0 & 0 & 1 \end{pmatrix} \\ &= \begin{pmatrix} r^2\sigma_\phi^2 \sin^2\phi & -r\sigma_\phi^2 \sin\phi\cos\phi & 0 \\ -r^2\sigma_\phi^2 \sin\phi\cos\phi & r^2\sigma_\phi^2 \cos^2\phi & 0 \\ 0 & 0 & \sigma_z^2 \end{pmatrix} \\ &= \begin{pmatrix} \sigma_\phi^2 y^2 & -\sigma_\phi^2 xy & 0 \\ -\sigma_\phi^2 xy & \sigma_\phi^2 x^2 & 0 \\ 0 & 0 & \sigma_z^2 \end{pmatrix}. \end{aligned}$$

This is sensible. The error and correlations involving $z$ are not affected. The errors on $x$ and $y$ depend on where they are; thus for $\phi$ near 0, $y$ is small and $x$ is along the radial direction and thus has a small error. The correlation between the errors on $x$ and $y$ is $\pm 1$.

*Who shall his errors understand?*
*Lord, make me pure within*
*From hidden faults.*

*—Psalm 19*

## 4.4 SYSTEMATIC ERRORS

If a meter has a random error of, say 1%, then its reading will sometimes be too high and sometimes too low; this is a nuisance, and your results will be accordingly erratic. However, you can live with it, and if necessary you can take repeated measurements to improve the accuracy.

If your meter has a systematic error such that it *consistently* reads 1% too high, this is different in two respects.

In the first place, there is no point in repeating measurements and averaging them. Nothing will be gained. The net effects of *random* errors decrease in size (usually like $1/\sqrt{N}$) as the number of measurements increases, but *systematic errors* do not fall off at all. Repeated measurements share the systematic effect and are therefore not independent, so that the central limit theorem, in the form of equation 4.5, does not apply. This is a longwinded way of saying the obvious: there is nothing to be gained from repeating your mistakes.

The second difference comes from the effects of non-independence of measurements at different points. A systematic error affects them all together. If you are unaware of it then you have no way of telling from your data that the effect is present and your results are wrong. They are internally consistent: if you plot voltage against current through a resistor, say, you will still find a beautiful straight line that goes through the origin—unfortunately the slope will be wrong.

This is why people are so frightened of systematic erros, and most other textbooks avoid the subject altogether. You never know whether you have got them and can never be sure that you have not—like an insidious disease. Random errors announce their existence by messing up the data, just as a conventional disease announces itself by painful symptoms so that due treatment can be taken at an early stage. However, an experiment with large systematic errors can look perfectly healthy, with nice straight lines and good $\chi^2$ for its fits, and yet the result is complete rubbish.

The good news, however, is that, despite popular prejudices and superstitions, once you know what your systematic errors are, they can be handled with standard statistical methods.

### 4.4.1 Finding, Eliminating, and Evaluating Them

Systematic errors cover a spectrum from the mildly inconvenient to the downright catastrophic.

At the more benign end of the range are factors with known errors you explicitly use in analysing your data. Think carefully about the whole experiment. Pay particular attention to all the numbers you apply to the data measurements—calibrations, efficiencies, etc. Worry hard about the accuracy of anything applied to all your data, unless it is completely above suspicion like $\pi$ or $\sqrt{2}$, as it affects all your data points in the same way and is thus a systematic effect.

*Example Thermocouple measurements*

You calibrate a thermocouple by measuring the output voltages $V_1$ and $V_2$, using a voltmeter of known resolution, at two standard temperatures, $T_1$ and $T_2$ ($T_1 < T_2$). You then use it to measure the temperature in your experiment, knowing that the voltage is linearly proportional to the temperature over the range concerned: for a voltage V,

$$T = \frac{T_2 - T_1}{V_2 - V_1}(V - V_1) + T_1.$$

These temperature measurements have an obvious random error due to the random error on $V$—but what about the errors on $V_1$ and $V_2$? If, for example, the measurement of $V_1$ is on the high side, then all the temperature measurements will systematically tend to be too low. However, the resolution of the meter which measured $V_1$ and $V_2$ is given, so these errors are known.

Less friendly are the numbers applied where you can not exactly quantify the error. Here one is reduced to intelligent guesswork, or sometimes just guesswork. In such cases the 'systematic errors' may be untrustworthy, though often they are much smaller than some other, more quantifiable, effect, so this is not too serious.

*Example Backgrounds*

You are measuring the efficiencies of several Geiger counters using a $Pb^{210}$ source, labelled 10 $\mu$Ci. The efficiency is given by

$$\eta = \frac{\text{observed rate}}{\text{true rate}}.$$

Now, 10 $\mu$Ci is 370 000 disintegrations per second—but unfortunately you do not know when it was measured, and the half-life of $Pb^{210}$ is only 21 years, so the true rate may really be less than the number on the label.

You decide, by inspecting the condition of the source container and from what you know about the way things work in your lab, that it is most unlikely for the source to be more than 5 years old, and in this worst case the rate is $370\,000 \times 2^{-5/21} = 313\,710$ counts per second. You take the most likely value as being midway between the two, 341 855, and appeal to the fact that the variance of a uniform distribution is 1/12 its width to give the error as $(370\,000 - 313\,710)/\sqrt{12} = 16\,250$, and the error is thus 5%. You are not entirely happy with this, but it is the best you can do with the data available, and you comfort yourself with the thought that it is probably an overestimate.

More serious are implicitly applied constants that may escape your notice. Often there are checks you can do to satisfy yourself and others of the absence of systematic effects, like plots that should be linear, or go through the origin. Ingenuity and raw cunning are useful here, and one has to be mildly paranoid and never take anything on trust.

*Example Pressure measurements*

You are studying the behaviour of low pressure gases, measuring the pressure with a mercury manometer. The difference in height between the two arms of mercury gives the difference between the pressure in the vessel and atmospheric pressure, which is 760 mm—but is it? It depends on the weather and the height of the laboratory above sea level, and if possible you should measure it. If not, then you have to estimate how much it might differ from the standard value.

Having done all you can on your own, the next thing to do is to ask Grumpy. Most of us have a colleague whose chief talent lies in destructive criticism—ask him politely (it is usually a him) what he thinks of your experiment. (Do not despair at the response! Before you decide to drop science and take up knitting instead, have a stiff drink and a good night's sleep, after which his critique of your experiment will seem less devastating.)

Naturally, such consultation is most effective at the planning stage of an experiment, as foreseen systematic effects can be countered by suitable design, or arranged to be monitored by suitable measurements.

The most vicious systematic errors are the ones you did not anticipate, such as an electronics component failing halfway through your experiment. You have two ways to defend yourself against this aspect of the basic hostility of things. The first is the use of consistency checks, if necessary repeatedly during the course of the experiment. The second is to randomise the order in which data are taken. Various nasty effects (electronics drifts, temperature drifts, and even psychological changes in the experimenter) are largely functions of time. If you take your data in an orderly sequence then this will become a systematic effect. Chopping and changing the order does not destroy these effects, but it renders them random rather than systematic.

The correct procedure depends on what you are trying to measure. This is very important in dealing with hysteresis effects in the apparatus (e.g. from adjustable controls with backlash). Measuring or setting the value of a quantity from below generally gives a different result than when approached from above. Thus, if the absolute values are important, such adjustments should be alternately done from above and from below, which will convert the systematic effect into a random one. On the other hand, if you are interested in a slope, i.e. the differences between two values is what matters, then the adjustments should all be made from the same side, as the systematic effects will cancel.

Having eliminated what systematic effects you can, you have to evaluate the ones you cannot. Often a systematic error is known exactly—for instance when a common calibration, with an unavoidable uncertainty, has been applied to all the values. Sometimes it is more nebulous and you have to make an intelligent guess.

Different independent systematic errors are, by virtue of their independence, added in quadrature. One or two are usually bigger than the rest and thus dominate the total value. This is often useful, for although you do have to work hard evaluating the large contributions, you do not have to sweat blood over the smaller ones.

If your final result has a random error and a systematic error, then, because the random and systematic errors are independent of each other, it follows directly from the central limit theorem that the total error is validly given by adding the two in quadrature. Even so, it is often helpful and useful to others to quote the two errors separately, and you will often see a statement like

$$A = -10.2 \pm 1.2 \pm 2.3$$

where the first error is statistical and the second systematic.

One reason for displaying results with separate errors in this way is that the statistical, random, error is usually more reliably known than the systematic, which tends (in some cases) to be an overestimate. It is also essential when combining or comparing experiments which share systematic effects. (It also shows that the first result is dominated by systematics, so that there is nothing to be gained by taking more data.)

### ★4.4.2 Living with Systematic Errors

Once you have found and estimated your systematic erros they become comparatively tame. They can be dealt with completely using the techniques of the covariance matrix, as described in section 4.3.4, which describes the way systematic errors affect the data points, and when you have tied down the errors in their covariance matrix, then you can use them in the combination of errors formula, equation 4.11 or 4.19, in the usual way.

Suppose two measurements $x_1$ and $x_2$ have a common systematic error $S$, and also individual random errors $\sigma_1$ and $\sigma_2$. This can be treated by considering $x_1$ as having two parts, $x_1^R$ with random error $\sigma_1$ and $x_1^S$ with systematic error $S$, and similarly $x_2$ is divided into $x_2^R$ and $x_2^S$. By this definition, $x_1^R$ and $x_2^R$ are independent of each other and of $x_1^S$ and $x_2^S$, whereas $x_1^S$ and $x_2^S$ are absolutely correlated. The variance of $x_1$ is given by

$$
\begin{aligned}
V(x_1) &= \langle x_1^2 \rangle - \langle x_1 \rangle^2 \\
&= \langle (x_1^R + x_1^S)^2 \rangle - \langle x_1^R + x_1^S \rangle^2 \\
&= \sigma_1^2 + S^2
\end{aligned}
$$

which incidentally justifies the statement in the previous section that systematic and random errors can be added in quadrature. Similar treatment gives $V(x_2) = \sigma_2^2 + S^2$ and the covariance

$$
\begin{aligned}
\operatorname{cov}(x_1, x_2) &= \langle x_1 x_2 \rangle - \langle x_1 \rangle \langle x_2 \rangle \\
&= \langle (x_1^R + x_1^S)(x_2^R + x_2^S) \rangle - \langle x_1^R + x_1^S \rangle \langle x_2^R + x_2^S \rangle .
\end{aligned}
$$

Three of the four cross products involve the $x^R$ and therefore cancel, as the $x^R$ are independent of everything else. The fourth involves $x_1^S$ and $x_2^S$, which are absolutely correlated, leaving us with

$$\operatorname{cov}(x_1, x_2) = \operatorname{cov}(x_1^S, x_2^S) = S^2.$$

Thus the variance matrix for $x_1$ and $x_2$ has the random and systematic errors added in quadrature along the diagonal; off the diagonal the covariances are given by the squared systematic error:

$$\mathbf{V} = \begin{pmatrix} \sigma_1^2 + S^2 & S^2 \\ S^2 & \sigma_2^2 + S^2 \end{pmatrix}. \tag{4.20}$$

So far we have considered systematic erros which are constants. They also appear as fractions or percentages; the systematic error $S$ is then not a constant but proportional to the measurement (or, more strictly, to the true value, but for 'small' errors the difference is by definition negligible):

$$S = \varepsilon x$$

with, for example, $\varepsilon = 0.01$ for a 1% error. This makes no difference to the above analysis. $x_1^S$ and $x_2^S$ are still absolutely correlated, and the error matrix is

$$\begin{pmatrix} \sigma_1^2 + \varepsilon^2 x_1^2 & \varepsilon^2 x_1 x_2 \\ \varepsilon^2 x_1 x_2 & \sigma_2^2 + \varepsilon^2 x_2^2 \end{pmatrix}.$$

This can all be generalised in the obvious way. If there are several independent sources of systematic error then they are added in quadrature. If there are more variables the matrix is larger. For example, consider three variables, $x_1, x_2, x_3$, and suppose that in addition to their random errors $\sigma_1$, $\sigma_2$, $\sigma_3$ they have a common systematic error $S$ and also another independent systematic error $T$ which is shared by $x_1$ and $x_2$ but not $x_3$. The covariance matrix is then (as you can show for yourself by multiplying out the covariances)

$$\begin{pmatrix} \sigma_1^2 + S^2 + T^2 & S^2 + T^2 & S^2 \\ S^2 + T^2 & \sigma_2^2 + S^2 + T^2 & S^2 \\ S^2 & S^2 & \sigma_3^2 + S^2 \end{pmatrix}.$$

*Example Ohm's law*

A current $I$ is determined by measuring the voltage $V$, using a meter of resolution $\sigma_V$, across a standard resistance $R \pm \sigma_R$. The error on $I$ is given by (see equation 4.15)

$$\frac{\sigma_I^2}{I^2} = \frac{\sigma_V^2}{V^2} + \frac{\sigma_R^2}{R^2}.$$

So far all is straightforward. However, if the apparatus is used to measure two currents, $I_1$ and $I_2$, the voltages $V_1$ and $V_2$ have independent errors $\sigma_V$, but the resistance is the same for both, and the error on it is thus systematic. We therefore rename it $S_R$ to denote its systematic status.

The errors on the measurements $I_1$ and $I_2$ are merely given by the above formula: nothing has changed there. However, there is a non-zero covariance between the two currents given by (equation 4.18).

$$\text{cov}(I_1, I_2) = \frac{\partial I_1}{\partial R}\frac{\partial I_2}{\partial R} S_R^2 = \frac{I_1 I_2}{R^2} S_R^2.$$

This matters if functions are formed from the two measurements. For example, using 4.18, the variance on $I_1 - I_2$ is $[2\sigma_V^2 + (I_1 - I_2)^2 \sigma_R^2]/R^2$. The case where the voltage measurement also contains a systematic effect is considered in the problems section at the end of this chapter.

## 4.5 PROBLEMS

*4.1*

Which is preferable, a set of ten measurements with a resolution of 1 mm or one measurement with a resolution of 0.2 mm?

*4.2*

Find the best combined result and error from the three measurements of $c$:

$$299\,798\,000 \pm 5000\ \text{m/sec}$$
$$299\,789\,000 \pm 4000\ \text{m/sec}$$
$$299\,797\,000 \pm 8000\ \text{m/sec}$$

*4.3*

Find the best combined result and error from the 5 measurements of $c$:

$$299\,794\,000 \pm 3000\ \text{m/sec}$$
$$299\,791\,000 \pm 5000\ \text{m/sec}$$
$$299\,770\,000 \pm 2000\ \text{m/sec}$$
$$299\,789\,000 \pm 3000\ \text{m/sec}$$
$$299\,790\,000 \pm 4000\ \text{m/sec}$$

*4.4*

If a voltage is determined by measuring a current of $1120 \pm 10$ mA through a resistance of $1400 \pm 30\,\Omega$, what is its value and error?

*4.5*

If a current is determined by measuring a current of $45 \pm 1$V through a resistance of $900 \pm 10\,\Omega$, what is its value and error?

*4.6*

If $\theta = 0.56 \pm 0.01$, what are the errors on $\sin\theta$, $\cos\theta$, and $\tan\theta$? What are they if $\theta = 1.56 \pm 0.01$?

*4.7*

Derive equation 4.7 from equation 4.6, using the combination of errors formula.

*4.8*

A (long-lived) source gives 389 counts in the first minute and 423 in the second minute. What is the best combined result? (Note: it is not 405.3.)

*4.9*

In the light of the previous problem, discuss the following exam question: 'Determine the best value and uncertainty for the strength of a particular radioactive source from the following set of independent measurements:

$$1.08 \pm 0.13 \quad 1.04 \pm 0.07 \quad 1.13 \pm 0.10 \quad \mu C_i\text{'}$$

★*4.10*

In the example on Ohm's law in section 4.4.2, suppose the error on the voltage comprises a systematic part $S_V$ and a random part $\sigma_V$. Find the covariance matrix for the three parameters $V_1$, $V_2$, $R$, the matrix $\mathbf{G}$, and thus find the error matrix for two current measurements.

CHAPTER 

# Estimation

In everyday life, 'estimation' means a rough and imprecise procedure leading to a rough and imprecise result. You 'estimate' when you cannot measure exactly.

In statistics, on the other hand, 'estimation' is a technical term. It means a precise and accurate procedure, leading to a result which may be imprecise, but where at least the extent of the imprecision is known. It has nothing to do with approximation. You have some data, from which you want to draw conclusions and produce a 'best' value for some particular numerical quantity (or perhaps for several quantities), and you probably also want to know how reliable this value is, i.e. what the error is on your estimate. This chapter deals first with the general problem of estimation and then with some specific examples and methods. One particular method, that of *least squares estimation*, is so important that it receives special treatment, and is dealt with in the next chapter.

## 5.1 PROPERTIES OF ESTIMATORS

**An estimator** *is a procedure applied to the data sample which gives a numerical value for a property of the parent population or, as appropriate, a property or parameter of the parent distribution function.*

This is purposefully drawn as a very general definition. The sample you have to study may have been drawn from some large *parent population*, for which you want to estimate some property. On the other hand, it may have been generated from a distribution function, arising from some basic law which you are investigating. This function also has properties; in addition it has parameters (like $p$ for a binomial) and you may be trying to measure one of these. Of course, in many cases a parameter is also a property (e.g. $\mu$ and $\sigma$ for a Gaussian).

Suppose you want to find the average height of all students in a university on the basis of an (honestly selected) sample of $N$. Here are some ways of getting the result, all fulfilling the above definition:

1. Add up all the heights and divide by $N$.
2. Add up the first 10 heights and divide by 10. Ignore the rest.
3. Add up all the heights and divide by $N - 1$.
4. Throw away the data and give the answer as 1.8 meters.
5. Multiply all the heights and take the $N$th root.
6. Choose the most popular height (the mode).
7. Add the tallest and shortest heights and divide by 2.
8. Add up the second, fourth, sixth, etc., heights and divide by $N/2$ for $N$ even, $(N - 1)/2$ for $N$ odd.

Each of these—yes, even 3 or 4—is an estimator, in that they each satisfy the above definition. Yet some are obviously better than others. Which of these would you use? Which do you regard as obviously stupid? An estimator cannot be described as 'right' or 'wrong', or as 'valid' or 'invalid', but it can be described as 'good' or 'bad'. Specifically, a 'good' estimator is *consistent*, *unbiased*, and *efficient*. A 'bad' estimator is *inconsistent*, *biased*, and *inefficient*. The meanings of these terms are described in the next section.

### 5.1.1 Consistency, Bias, and Efficiency

Suppose that the quantity we are trying to measure is called $a$. We will use the symbol ^ for estimators, so $\hat{a}$ denotes an estimator for $a$.

When an estimator $\hat{a}$ is applied to the $N$ measurements of some particular data simple, it produces an estimate of the quantity $a$. This can and will differ from the true value, because of the effects of statistical fluctuations in the sample. However, we know from the law of large numbers (section 3.1.2) that these effects get smaller and vanish as $N$ increases to infinity, provided the measurements are independent, i.e. there are no systematic errors. So it seems a reasonable thing to demand of a 'good' estimator that the difference between the estimate and the true value vanish for large samples. Such an estimator is called *consistent*.

**An estimator is consistent** *if it tends to the true value as the number of data values tends to infinity*:

$$\lim_{N \to \infty} \hat{a} = a.$$

In the previous examples of estimators for the mean height, $\mu$, it is easy to show that 1 is consistent:

$$\hat{\mu} = \frac{x_1 + x_2 + x_3 + \cdots + x_N}{N} = \bar{x}$$

and as $N \to \infty$, the law of large numbers guarantees that $\bar{x} \to \mu$; 3 is also consistent, as the difference between $N-1$ and $N$ vanishes for large $N$. Similarly, 8 is consistent. On the other hand, 2 and 4 are obviously inconsistent. This gives us a reason for not using them, but does not enable us to say whether 1, 3, or 8 is preferable.

For a finite $N$ we cannot hope that, for a particular sample of data, $\hat{a}$ will have the same value as the true $a$. It may be too large or too small. However, we can reasonably require that the chances of an overestimate balance those of an underestimate. Such an estimator is called *unbiased*.

**An estimator is unbiased** *if its expectation value is equal to the true value*:

$$\langle \hat{a} \rangle = a.$$

Now look at (1) again. Its expectation value is

$$\begin{aligned}
\langle \hat{\mu} \rangle &= \left\langle \frac{x_1 + x_2 + x_3 + \cdots + x_N}{N} \right\rangle \\
&= \frac{\langle x \rangle + \langle x \rangle + \langle x \rangle + \cdots + \langle x \rangle}{N} \\
&= \frac{N \langle x \rangle}{N} \\
&= \mu.
\end{aligned}$$

So (1) is unbiased. However, for 3 the calculation gives

$$\langle \hat{\mu} \rangle = \frac{N}{N-1} \mu \neq \mu$$

showing that 3 is biased. So we are justified in using 1 rather than 3. On the other hand, 8 passes this test, as does 2.

The value of $\hat{a}$ depends on the data sample, and the particular data sample you happen to obtain is a matter of chance. Depending on the data you chance to obtain, the value of $\hat{a}$ will vary. If the estimate is to be a good measurement of the true value, we would like this spread of possible values to be as small as possible. An estimator with a small spread, as measured by its variance, is called *efficient*.

**An estimator is efficient** *if its variance is small.*

If the variance of one estimator is smaller than that of another, it is more likely to be close to the true value, and is therefore better. Thus we can say that as 8 above uses only half the data that 1 uses, but is otherwise similar, its variance is larger by a factor of $\sqrt{2}$. Therefore 8 is less efficient than 1, and this justifies using 1 rather than 8. This is an artificially blatant example; for real estimators the difference is usually less obvious, but the principle is the same.

The choice of estimator to use in a particular application requires judgement: there is no such thing as an ideal 'best' estimator. There are two reasons for this. One is that the variance of the estimator depends on the distribution concerned, so the efficiency of an estimator may be different for different problems. (This is illustrated in section 5.1.2.) The other reason is that detailed analysis may show that the most efficient estimator is biased, so you have to weigh the relative merits of an unbiased, inefficient estimator against another which is more efficient but slightly biased.

### ★5.1.2 The Likelihood Function

When confronted with a particular data sample, $\{x_1, x_2, x_3, \ldots, x_N\}$, one applies $\hat{a}$, an estimator for the quantity $a$, and obtains a particular numerical value, which is the desired estimate. One hopes that this $\hat{a}(x_1, \ldots, x_N)$ is close to the real value of $a$, though the actual value will depend on the particular set of data values $\{x_i\}$.

When considering the properties of estimators, one starts from the opposite end. Instead of looking at a particular sample, one considers a known distribution. The data values $x_i$ are drawn from some probability density function $P(x; a)$ which depends on $a$. The form of the function $P$ is given and $a$ is specified. (We assume, for the purpose of illustration, that $a$ is a parameter rather than a property.) $\hat{a}$ is a function of the $x_i$ and as such has an expectation value, so one can say things like: 'If I take a sample of $N$ values from this distribution, and from them form $\hat{a}(x_1, x_2, \ldots, x_N)$, then on average (over many such samples) the value of $\hat{a}$ would be $\langle \hat{a} \rangle$.'

The probability of a particular set of data $\{x_1, x_2, x_3, \ldots, x_N\}$ is the product of the individual probabilities. This product is called the *likelihood* $L(x_1, x_2, x_3, \ldots, x_N; a)$. It is the combined probability, or probability density, that this particular set of $x_i$ would be produced from this value of $a$:

$$\begin{aligned} L(x_1, x_2, \ldots, x_N; a) &= P(x_1; a)P(x_2; a) \cdots P(x_N; a) \\ &= \prod P(x_i; a). \end{aligned} \tag{5.1}$$

The expectation value for any function of the $\{x_i\}$ is found by integrating over all possible values of all the $x_i$, weighting by the total probability, i.e. the likelihood:

$$\langle f(x_1,\ldots,x_N)\rangle = \int\cdots\int f(x_1,\ldots,x_N)L(x_1,\ldots,x_N;a)\,\mathrm{d}x_1,\ldots,\mathrm{d}x_N.$$

This looks horribly complicated; let us conceal it under the obvious notation

$$\langle f(x_1,\ldots,x_N)\rangle = \int fL\,\mathrm{d}X \tag{5.2}$$

(but one has to remember that it is a function of $a$). In particular

$$\langle \hat{a}\rangle = \int \hat{a}L\,\mathrm{d}X \qquad \langle \hat{a}^2\rangle = \int \hat{a}^2L\,\mathrm{d}X. \tag{5.3}$$

In this language, the consistency requirement is that

$$\lim_{N\to\infty} \langle \hat{a}-a\rangle = 0.$$

It is not usually difficult to find a consistent estimator, as the law of large numbers is on your side. Even so, this cannot be taken for granted. We saw in the previous section that the estimators 2 and 4 were inconsistent, though these were obviously not meant as serious suggestions. Some estimators can be shown to be always consistent, like 1, 3, and 8. Others are case-dependent. Consider 7, for example. This is consistent if the distribution is symmetric about $x=\mu$ (for reasons of symmetry), but not if it is asymmetric. Likewise 6 is consistent if, and only if, the mode and mean of the distribution are the same.

Bias is similar. Some estimators are guaranteed bias-free under all circumstances. Some are inescapably biased. Others may or may not be biased, depending on the distribution concerned.

If a bias is found, it is easy to correct for it. If $\langle \hat{a}\rangle = b$, then $(\hat{a}-b)$ is bias-free. Incidentally, if an estimator is consistent it follows that its bias (if any) vanishes as $N\to\infty$.

There is no such thing as a generally efficient estimator. Efficiency depends on the case considered, i.e. the form of $P(x;a)$ and possibly also the value of $a$. For a given function $P(x;a)$ the function $\hat{a}$ has a probability distribution. An 'efficient' estimator is one for which the spread of this distribution about its mean is small. This is described by the variance $V(\hat{a}) = \langle(\hat{a}-a)^2\rangle = \langle \hat{a}^2\rangle - \langle \hat{a}\rangle^2$ which can be found from the likelihood integrals (equation 5.3). If you want to consider the merits of two rival estimators, $\hat{a}_1$ and $\hat{a}_2$, you can do so *provided* you know the expression for $P(x;a)$ and are prepared to wade through a lot of integration.

There is a surprisingly powerful result (derived in the next section) which says that there is a limit to the accuracy of an estimator. This is called the *minimum variance bound*, or MVB for short.

For an unbiassed estimator the MVB is

$$V(\hat{a}) \geqslant \frac{1}{\langle (\mathrm{d}\ln L/\mathrm{d}a)^2 \rangle} \tag{5.4}$$

which can also be written

$$V(\hat{a}) \geqslant \frac{-1}{\langle (\mathrm{d}^2 \ln L/\mathrm{d}a^2) \rangle} \tag{5.5}$$

where $L$ is the likelihood function, defined in equation 5.1.

If, for some estimator $\hat{a}$, $V(\hat{a})$ is equal to the MVB, then $\hat{a}$ is said to be 'efficient'. If not, its 'efficiency' is MVB$/V(\hat{a})$.

For a Gaussian distribution, the sample mean is an efficient estimate of $\mu$.

The probability of a particular $x_i$ is $p(x_i; \mu) = \dfrac{1}{\sigma\sqrt{2\pi}} \mathrm{e}^{-(x_i-\mu)^2/2\sigma^2}$

giving a total log likelihood $\ln L = -\sum_i \dfrac{(x_i-\mu)^2}{2\sigma^2} - N\ln(\sigma\sqrt{2\pi})$.

Differentiating twice with respect to $\mu$ gives $\dfrac{\mathrm{d}^2 \ln L}{\mathrm{d}\mu^2} = -\dfrac{N}{\sigma^2}$.

This does not involve $x$ so this expression is equal to its expectation value without further manipulation. Inverting and negating gives

$$\mathrm{MVB} = \frac{\sigma^2}{N}$$

which is also (by the central limit theorem) the variance of the estimator.

*Example A uniform distribution*

Consider by contrast the estimation of the mean of a distribution which is guaranteed to be uniform between two (unknown) limits. The sample mean provides a consistent and unbiased estimator, but it is not efficient.

This makes sense: suppose you had measured four values, at 1.1, 1.3, 1.7, and 1.6 (for example). A fifth measurement at 1.4 will modify the mean—and yet it gives no new useful information. The two extreme measurements are the only ones of any real significance. Using them to give the midrange

$$\hat{\mu} = \tfrac{1}{2}[\min(x_i) + \max(x_i)]$$

gives an estimator which for a uniform distribution of width $W$ has variance $W/2(N+1)(N+2)$ (this can be shown by working out the probability distributions for $\min(x)$ and $\max(x)$). The variance of the mean, on the other hand, is still $\sigma^2/N$, and (using equation 3.24) this is $W/12N$. So in this situation the variance of the midrange is less than the variance of the mean (provided $N$ is greater than 1).

This is also interesting in that, for large $N$, the variance falls like $1/N^2$, and the standard deviation falls like $1/N$, instead of the usual $1/\sqrt{N}$ behaviour of random errors.

### ★5.1.3 Proof of the Minimum Variance Bound

The statement that $a$ is unbiased means that

$$\langle \hat{a} \rangle = \int \hat{a} L \, \mathrm{d}X = a.$$

Differentiate this with respect to $a$. $\hat{a}$ depends only on the $x_i$, so the differential only affects $L$, giving

$$\int \hat{a} \frac{\mathrm{d}L}{\mathrm{d}a} \mathrm{d}X = 1$$

provided that the limits of the $x$ integrations do not depend on $a$. This can be rewritten as

$$\int \hat{a} \frac{\mathrm{d} \ln L}{\mathrm{d}a} L \, \mathrm{d}X = 1.$$

Now, the total probability must be normalised, i.e.

$$\int L \, \mathrm{d}X = 1.$$

Differentiating this and treating in the same way gives

$$\int \frac{\mathrm{d} \ln L}{\mathrm{d}a} L \, \mathrm{d}X = 0.$$

This can also be written as

$$\left\langle \frac{\mathrm{d} \ln L}{\mathrm{d}a} \right\rangle = 0. \tag{5.6}$$

Multiply this by $a$ and subtract it from the previous result to get

$$\int (\hat{a} - a) \frac{\mathrm{d} \ln L}{\mathrm{d}a} L \, \mathrm{d}X = 1.$$

Now invoke the Schwarz inequality †

$$\int u^2 \, \mathrm{d}X \int v^2 \, \mathrm{d}X \geqslant \left( \int uv \, \mathrm{d}X \right)^2 \tag{5.7}$$

†To see this, note that $\int (\lambda u + v)^2 \, \mathrm{d}X$ is positive (or zero) whatever the value of $\lambda$. Hence $\lambda^2 \int u^2 \, \mathrm{d}X + 2\lambda \int uv \, \mathrm{d}X + \int v^2 \, \mathrm{d}X$ has no real non-zero roots for $\lambda$. Now apply the quadratic formula.

with

$$u \equiv (\hat{a} - a)\sqrt{L} \qquad v \equiv \frac{\mathrm{d}\ln L}{\mathrm{d}a}\sqrt{L}.$$

This gives

$$\left(\int (\hat{a} - a)^2 L\,\mathrm{d}X\right)\left(\int \left(\frac{\mathrm{d}\ln L}{\mathrm{d}a}\right)^2 L\,\mathrm{d}X\right) \geqslant 1.$$

The first integral is $\langle(\hat{a}-a)^2\rangle$, the variance of the estimator. The second is the expectation value of the squared differential of the logarithm of the likelihood, so

$$V(\hat{a})\left\langle\left(\frac{\mathrm{d}\ln L}{\mathrm{d}a}\right)^2\right\rangle \geqslant 1$$

giving equation 5.4:

$$V(\hat{a}) \geqslant \frac{1}{\langle(\mathrm{d}\ln L/\mathrm{d}a)^2\rangle}.$$

The quantity in the denominator is sometimes called the *information*, $I(a)$ (following R.A. Fisher). It can be rewritten in various forms. Differentiating equation 5.6 again gives

$$\int \left(\frac{\mathrm{d}^2\ln L}{\mathrm{d}a^2}L + \frac{\mathrm{d}\ln L}{\mathrm{d}a}\frac{\mathrm{d}L}{\mathrm{d}a}\right)\mathrm{d}X = 0$$

and rearrangement produces

$$\left\langle\frac{\mathrm{d}^2\ln L}{\mathrm{d}a^2}\right\rangle = -\left\langle\left(\frac{\mathrm{d}\ln L}{\mathrm{d}a}\right)^2\right\rangle \tag{5.8}$$

giving the form for equation 5.5. It can also be written using the individual probability function $P(x;a)$:

$$I(a) = -N\int \frac{\mathrm{d}^2\ln P}{\mathrm{d}a^2}P\,\mathrm{d}x = N\int\left(\frac{\mathrm{d}\ln P}{\mathrm{d}a}\right)^2 P\,\mathrm{d}x. \tag{5.9}$$

Note that as $L$ is, presumably, at or near a maximum, its second derivative is negative, and that is why the minus sign comes in.

For a biased estimator, if the bias is $b$, then the numerator in the MVB is replaced by $(1 + \mathrm{d}b/\mathrm{d}a)^2$.

The bound is also known as the Cramér-Rao inequality and also as the Fréchet inequality. It was first discovered by Aitken and Silverstone.

## 5.2 SOME BASIC ESTIMATORS

The previous section describes various general properties of estimators. In this section some specific estimators are described, and you can see how these ideas apply in actual cases.

### 5.2.1 Estimating the Mean

The mean of a sample provides an estimate of the true mean

$$\hat{\mu} = \bar{x}. \tag{5.10}$$

As is shown in section 5.1.1, the central limit theorem (see section 4.1) guarantees that this estimate is consistent and unbiased, and that its variance is

$$V(\hat{\mu}) = \frac{\sigma^2}{N} \tag{5.11}$$

where $N$ is the number of data values and $\sigma$ is the standard deviation of the parent distribution.

This is also discussed in section 4.2.1, under the heading of 'repeated measurements' where the idea of using the arithmetic mean of a sample as a 'measurement' (i.e. estimate) of the true mean occurs naturally, without the need for technical discussion of 'consistency', 'bias', or whatever. The difference between the standard deviation $\sigma$, describing the extent to which any single measurement is liable to vary from the mean $\mu$, and the 'standard error on the mean', $\sigma/\sqrt{N}$—the square root of equation 5.11—describing the extent to which your estimate is liable to differ from the true mean, is emphasised in that section; no apologies are given for stressing it again here, as the difference between these is very important and confusion between them is a common mistake.

This estimator may be efficient or it may not. There may be no other estimator with a smaller variance, or again there may be. It depends on the parent distribution. For a Gaussian distribution it is indeed efficient—proof, if required, is given in section 5.1.2. However, there are other distributions (see section 5.1.2 again for an example) for which you can do better.

### 5.2.2 Estimating the Variance

Consider first the case where the true mean $\mu$ is known. (This is actually rather an unusual case, but it is simpler to consider it first, before the more usual case of unknown $\mu$.) An obvious estimator to use is

$$\widehat{V(x)} = \frac{1}{N} \sum (x_i - \mu)^2. \tag{5.12}$$

It is easy to see that this is consistent, and that it is unbiased as

$$\langle \widehat{V(x)} \rangle = \frac{N\langle (x-\mu)^2 \rangle}{N} = \langle (x-\mu)^2 \rangle = V(x).$$

Now suppose that we do not know $\mu$. An obvious remedy is to replace $\mu$ in equation 5.12 by our estimate $\hat{\mu} = \bar{x}$:

$$\widehat{V(x)} = \frac{1}{N}\sum(x_i - \bar{x})^2 = \frac{1}{N}\sum(x_i^2 - \bar{x}^2). \tag{5.13}$$

However, this is biased. When you take expectation values you get

$$\begin{aligned}\langle \widehat{V(x)} \rangle &= N\frac{\langle x^2 - \bar{x}^2 \rangle}{N} \\ &= \langle x^2 \rangle - \langle \bar{x}^2 \rangle\end{aligned}$$

As $\langle x \rangle = \langle \bar{x} \rangle$, from the CLT (section 4.1), this is

$$\begin{aligned}\langle \widehat{V(x)} \rangle &= \langle x^2 \rangle - \langle x \rangle^2 - (\langle \bar{x}^2 \rangle - \langle \bar{x} \rangle^2) \\ &= V(x) - V(\bar{x})\end{aligned}$$

and the CLT also gives $V(\bar{x}) = V(x)/N$, so

$$\begin{aligned}\langle \widehat{V(x)} \rangle &= \left(1 - \frac{1}{N}\right)V(x) \\ &= \frac{N-1}{N}V(x) \neq V(x).\end{aligned}$$

So this gives us an estimate of the parent variance which is biased (though the bias falls like $1/N$ and can be neglected if $N$ is comfortably large). This is understandable. $V(x)$ involves the squared deviation from the true mean $\mu$. Instead we taken it from the estimated mean $\bar{x}$. Now, the quantity $\sum(x_i - \xi)^2$ has a minimum at $\xi = \bar{x}$, so the estimated variance, as defined by equation 5.13, must be less than, or, at best, equal to, the unbiased estimate of equation 5.12.

Having found that the bias is there, it is easy to correct for it. Multiplying by $N/(N-1)$, known as *Bessel's correction*, exactly compensates for the lack of bias. Thus

$$\widehat{V(x)} = s^2 = \frac{1}{N-1}\sum(x_i - \bar{x})^2 \tag{5.14}$$

is consistent and bias-free. The symbol $s$ is introduced here because of the importance of the formula: $s^2$ is the *unbiased* estimate of the variance. Note that this estimator does not work for $N = 1$, which makes sense, if you think about it.

Now we have to evaluate the variance of this estimator. This can be done by starting from the general case of the $r$th moment $\overline{x^r}$. This has variance

$$V(\overline{x^r}) = \frac{1}{N}\int(x^r - \langle x^r \rangle)^2 P(x)\,dx = \frac{1}{N}(\langle x^{2r} \rangle - \langle x^r \rangle^2) \tag{5.15}$$

and likewise

$$\operatorname{cov}(\overline{x^r}, \overline{x^q}) = (\langle x^{r+q} \rangle - \langle x^r \rangle \langle x^q \rangle)/N. \tag{5.16}$$

The estimator of equation 5.13 is not the second moment, but the second central moment (see section 2.5.2). Its variance is more complicated than that of 5.15, because the constant $\langle x \rangle$ is replaced by the variable $\bar{x}$. Writing down $\langle \widehat{V(x)}^2 \rangle - \langle \widehat{V(x)} \rangle^2$ and reducing this to expectation values of the moments by writing out the products of the summations eventually gives, with the help of 5.15 and 5.16 (writing $\mu$ in place of $\langle x \rangle$, for clarity),

$$V(\widehat{V(x)}) = \frac{(N-1)^2}{N^3} \langle (x-\mu)^4 \rangle - \frac{(N-1)(N-3)}{N^3} \langle (x-\mu)^2 \rangle^2.$$

If $N$ is large—and this is usually a fair approximation—this reduces to

$$V(\widehat{V(x)}) = \frac{1}{N} [\langle (x-\mu)^4 \rangle - \langle (x-\mu)^2 \rangle^2]. \tag{5.17}$$

For a Gaussian distribution, using the integrals of Table 3.1, this reduces to

$$V(\widehat{V(x)}) = \frac{2\sigma^4}{N}. \tag{5.18}$$

If Bessel's correction is applied, i.e. if equation 5.14 is used instead of 5.13, the factor of $N$ in the denominator is replaced by $N-1$.

### 5.2.3 Estimating $\sigma$

The obvious way of estimating $\sigma$ is just to take the square root of the estimate for the variance, equation 5.12, 5.13, or 5.14 as applicable:

$$\hat{\sigma} = \sqrt{\widehat{V(x)}} \tag{5.19}$$

or, even more briefly, $\hat{\sigma}^2 = \widehat{\sigma^2}$. For equation 5.14 this gives

$$\hat{\sigma} = s. \tag{5.20}$$

The law of large numbers guarantees consistency as before. The situation as regards bias is less happy: although equation 5.14 is an unbiased estimate of $V(x)$, and thus of $\sigma^2$, this does not mean that its square root is an unbiased estimate of $\sigma$. However, this does not matter unduly as in calculations of errors or significance the standard deviation appears as $\sigma^2$.

The variance of $\hat{\sigma}$ can be found from equation 4.9:

$$V(\hat{\sigma}^2) = \left( \frac{d\sigma^2}{d\sigma} \right)^2 \qquad V(\hat{\sigma}) = 4\sigma^2 V(\hat{\sigma}).$$

Inserting equation 5.17 gives the variance of $\hat{\sigma}$, as defined by equations 5.13

and 5.19, if $N$ is reasonably large:

$$V(\hat{\sigma}) = \frac{\langle (x-\mu)^4 \rangle - \langle (x-\mu)^2 \rangle^2}{4N\sigma^2} \tag{5.21}$$

which for a Gaussian reduces to

$$V(\hat{\sigma}) = \frac{\sigma^2}{2N} \tag{5.22}$$

which can also be written

$$\sigma_\sigma = \frac{\sigma}{\sqrt{2N}}. \tag{5.23}$$

If the unbiased estimator $s$ (equation 5.14) is used instead, then this is increased to

$$\sigma_s = \frac{\sigma}{\sqrt{2(N-1)}}. \tag{5.24}$$

A little thought has to be given to the meaning and use of equations 5.21 to 5.24. On the face of it, equation 5.24 tells you the error of your estimate of $\sigma$, in terms of $\sigma$ itself. To evaluate $V(\hat{\sigma})$ or $\sigma_\sigma$ you need to know what $\sigma$ is, but in that case why should you be trying to estimate it? What use is a formula that only tells you something if you already know the answer?

Despite this, there are two ways of applying these formula sensibly. The first is at the 'planning' or 'design' stage of an experiment, when you have no data, but need to plan ahead and consider the results that might be obtained, for various actual true values. You use them to make statements like: 'I want to measure $\sigma$. If it has the value 6 then with 40 measurements I can measure it with an accuracy of 0.7.' You could use it to find how many measurements would be necessary to achieve some desired precision, given reasonable assumptions about the size of the true value.

The other situation where you apply these formulae is in handling the real experimental results. Now you do not know the real value of $\sigma$ to use in equation 5.23 or 5.24 or possibly even 5.11, but you do have an estimate $\hat{\sigma}$ which you can use instead. This gives you an estimate of the error which is presumably reasonably close to the truth. Thus if you measure a $\sigma$ of 6.0 from 40 measurements, known to be described by the Gaussian distribution, then you can legitimately quote the results as $6.0 \pm 0.7$. Even so, there is always a difference between an error obtained from the data in such a way, and one that is known *a priori*. Similarly, in using equation 5.11 for the variance of the mean, there is a difference between a quoted result of $\mu = 10.0 \pm 0.1$ coming from four measurements, each with a known error of 0.2, and a result of $\mu = 10.0 \pm 0.1$ coming from four measurements where the resolution has been estimated as 0.2 on the basis of equation 5.14. Such differences are important for small $N$, and require to be dealt with by the use of *Student's t distribution* rather the Gaussian for describing errors, as is described in Chapter 7. For large $N$ the estimates improve and the difference becomes unimportant.

In real measurements this limit, like any other large $N$ limit, tends to run into problems of systematic errors. Any physical measurement usually has a systematic error associated with it—the zeroing error on a meter, the internal resistance of a voltage source, or whatever. Sensible design should make such errors smaller than the random error on a single measurement, but if you reduce the random effects by taking a large number of measurements, the previously negligible systematic effects become dominant.

### ★5.2.4 The Correlation Coefficient

The correlation coefficient (see section 2.6) within a sample can be taken as an estimate of the correlation of the parent distribution

$$\hat{\rho} = \frac{\overline{xy} - \bar{x}\bar{y}}{\sigma_x \sigma_y}. \tag{5.25a}$$

Again, Bessel's correction can be used to remove bias:

$$\hat{\rho} = r = \frac{\overline{xy} - \bar{x}\bar{y}}{(N-1)s_x s_y}. \tag{5.25b}$$

In fact there is a somewhat more efficient estimator, but it involves the solution of a cubic equation (see Kendall and Stuart, Sect. 18.9).

When the size of the data sample, $N$, is *very* large it can be shown that the error on this is given by

$$\sigma_\rho = \frac{1-\rho^2}{\sqrt{N-1}}.$$

A better approximation, which works well down to moderate $N$, is to transform to the variable $z$:

$$z = \frac{1}{2}\ln\frac{1+\hat{\rho}}{1-\hat{\rho}}.$$

If $x$ and $y$ are distributed according to a (general two-dimensional) Gaussian, then $z$ has a more Gaussian shape than $\hat{\rho}$. The standard deviation for $z$ is $1/\sqrt{N-3}$.

*Example Illiterate scientists?*

Fifteen physics students were given an essay to write, which was then assessed. The correlation between the essay mark and their mark in the end-of-year exams was found to be $r = -0.11$. If this correlation were really negative it would imply that literate students are bad at physics, and vice versa. Is there any support for this?

Transforming from $r$ to $z$ we get

$$z = \frac{1}{2}\ln\left(\frac{0.89}{1.11}\right) = -0.1104.$$

The error is $1/\sqrt{12} = 0.2887$. So the deviation from zero is less than half a standard deviation and is not significant.

*I shall stick to the principle of likelihood which I laid down at the start, and try to give an account of everything from the beginning that will be more rather than less likely. So let us begin again, calling as we do on some protecting deity to see us safely through a strange and unusual arrangement to a likely conclusion.*

—*Plato:* Timaeus

## ★5.3 MAXIMUM LIKELIHOOD

The *principle of maximum likelihood* (ML for short) is a method for estimation. For a data sample $\{x_1, \ldots, x_N\}$, the maximum likelihood estimator $\hat{a}$ is the value of $a$ for which the likelihood in equation 5.1.

$$L(x_1, \ldots, x_N; a) = \prod P(x_i, a)$$

is a maximum. You determine the value of $a$ that makes the probability of the actual results obtained, $\{x_1, \ldots, x_N\}$, as large as it can possibly be. In practice it is easier to maximise the logarithm of $L$, which is the sum of the logarithms of the probabilities.

For an over-simple example, suppose that five values of $x$, namely

$$0.89, 0.03, 0.50, 0.36, 0.49$$

have been generated by a distribution which is known to have the form $P(x) = 1.0 + a(x - 0.5)$ between 0 and 1, and zero elsewhere. From these five values we can calculate the logarithm on the likelihood of various values of $a$, and plot the result (Figure 5.1). For example:

if $a = +1.0$ it is $\ln 1.39 + \ln 0.53 + \ln 1.0 + \ln 0.86 + \ln 0.99 = -0.47$
if $a = -1.0$ it is $\ln 0.61 + \ln 1.47 + \ln 1.0 + \ln 1.14 + \ln 1.01 = +0.03$.

The log likelihood for values of $a$ between $-1.0$ and 1.0, is shown. From it you can see that the maximum likelihood occurs at roughly $a = -0.6$, so this is the maximum likelihood estimate for $a$.

In many cases you do not find the maximum directly, by drawing pictures or otherwise. Instead it is easier to find the position of the maximum by

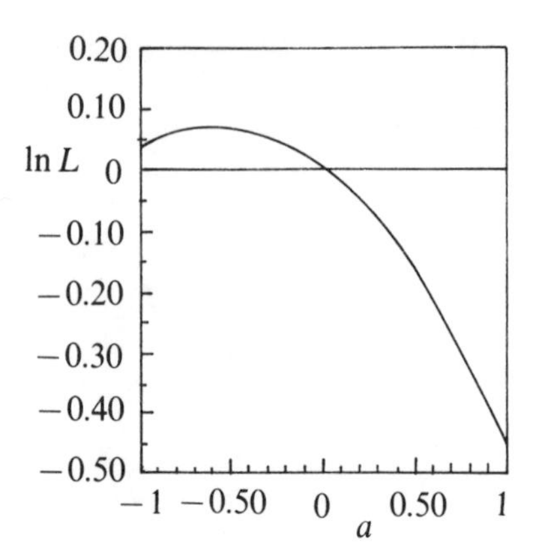


Fig. 5.1. A likelihood function.

differentiating the expression for $\ln L$ and setting it to zero:

$$\left.\frac{\mathrm{d}\ln L}{\mathrm{d}a}\right|_{a=\hat{a}} = 0. \tag{5.26}$$

Sometimes this gives you an equation that you can solve analytically to get a closed form for the estimate. If not, you have to solve it numerically.

*Example Simple lifetime*

Suppose you are studying the decay of some state of unknown lifetime, i.e. such states decay like $(1/\tau)\mathrm{e}^{-t/\tau}$, and you want to find $\tau$. You accumulate a set of $N$ observed lifetimes $\{t_i\}$. The log likelihood function is

$$\ln L = \sum \ln\left(\frac{1}{\tau}\mathrm{e}^{-t_i/\tau}\right) = \sum -\frac{t_i}{\tau} - \ln\tau.$$

Differentiating with respect to $\tau$ gives

$$\frac{\mathrm{d}\ln L}{\mathrm{d}\tau} = \sum\left(\frac{t_i}{\tau^2} - \frac{1}{\tau}\right).$$

Setting this to zero gives the maximum likelihood estimator, $\hat{\tau}$:

$$\sum\left(\frac{t_i}{\hat{\tau}^2} - \frac{1}{\hat{\tau}}\right) = 0$$

$$\hat{\tau} = \frac{1}{N}\sum t_i.$$

*Example Lifetime with acceptance*

To show the flexibility of the ML method, suppose you do an experiment similar to the last, except that your equipment fails to register any decays longer than some time $T$. The probability function is then (suitably normalised to 1)

$$\frac{1}{\tau}e^{-t/\tau}(1-e^{-T/\tau})^{-1}.$$

Differentiating the log likelihood obtained from this and setting it to zero gives

$$\hat{\tau}=\frac{1}{N}\sum t_i+\frac{1}{N}\sum\frac{Te^{-T/\hat{\tau}}}{(1-e^{-T/\hat{\tau}})}$$

which has to be solved numerically to give $\hat{\tau}$.

Several simple estimators of Gaussian properties that we met earlier can also be derived by maximum likelihood.

*Example ML and the Gaussian weighted mean*

Suppose $\{x_i\}$ are measurements of the same quantity, but with differing precision, i.e. $x_i$ is taken from a Gaussian of mean $\mu$ (which is what one is trying to find) and standard deviation $\sigma_i$ (which is known—it is the error of the measurement). The probability function for $x_i$ is

$$P(x_i;\mu,\sigma_i)=\frac{1}{\sigma_i\sqrt{2\pi}}e^{-(x-\mu)^2/2\sigma_i^2}$$

so the logarithm of the likelihood is

$$\ln L=\sum_i-\ln\sigma_i\sqrt{2\pi}-\sum_i\frac{(x_i-\mu)^2}{2\sigma_i^2}$$

and differentiating this with respect to $\mu$ to find the maximum gives

$$\sum\left(\frac{x_i-\hat{\mu}}{\sigma_i^2}\right)=0$$

$$\hat{\mu}=\frac{\sum(x_i/\sigma_i^2)}{\sum(1/\sigma_i^2)}$$

i.e. the $x_i$ values are averaged, each weighted by $1/\sigma_i^2$, and the answer appropriately normalised. This agrees with the earlier treatment of averaging weighted data, in section 4.2.2.

*Example ML estimate of Gaussian mean and σ*

Suppose that you have a set of measurements of a quantity $\mu$, where the resolution $\sigma$ is the same for all the $x_i$, but is unknown. The expression for $\ln L$ has therefore to be differentiated with respect to $\mu$ and with respect to $\sigma$, giving two equations. These are (try it!)

$$\sum(x_i-\hat{\mu})=0$$

$$\sum\frac{(x_i-\hat{\mu})^2}{\hat{\sigma}^3}-\sum\frac{1}{\hat{\sigma}}=0.$$

In principle these are two simultaneous equations to be solved, but in fact, rather conveniently, there is no $\hat{\sigma}$ appearing in the first equation, which gives $\hat{\mu} = \bar{x}$ as before. Having obtained the value for $\hat{\mu}$ this can be substituted in the second equation to give

$$\hat{\sigma}^2 = \frac{1}{N} \sum (x_i - \bar{x})^2.$$

The result of this last example is essentially just equation 5.13, which is known to be biased. Why this happens is discussed in the next section.

### ★5.3.1 ML: Consistency, Bias, and Invariance

ML estimators are usually (though not always) consistent. However they are, in general, biased. This is not mentioned every often, partly because the bias becomes small as the sample size gets reasonably large. Even so, it seems odd that such a much-used method of estimation should have this undesirable feature built into it, until you realise that bias is the price you have to pay for another desirable property of maximum likelihood estimates, namely the *invariance* of ML estimators under parameter transformations.

For instance, in the above example, instead of trying to estimate the standard deviation $\sigma$, we could have chosen to estimate the variance $\sigma^2$. If you do this, by differentiating $L$ with respect to $\sigma^2$ and setting it to zero, this gives the estimator for the variance

$$\widehat{\sigma^2} = \frac{1}{N} \sum (x_i - \hat{\mu})^2$$

which shows that the estimator of $(\sigma^2)$ and the (estimator of $\sigma)^2$ are the same. This is always true for ML estimators. We are looking for the maximum peak of the likelihood function $L$. If this occurs at some particular value $a = a_1$, then at this turning point $a^2 = a_1^2$, $\sqrt{a} = \sqrt{a_1}$, and so on: we could have worked with $a^2$ or $\sqrt{a}$ and would have found the same peak in the same place. In general if we prefer to estimate some function $f(a)$ rather than $a$, then

$$\widehat{f(a)} = f(\hat{a}).$$

This means that we do not have to worry about the distinction between them, which is very convenient. However, it can not be taken for granted. Other estimation procedures preserve this difference.

Invariance between two estimators is incompatible with lack of bias in both. Even if one is bias-free, any non-trivial transformation to another variable will give a biased estimate. The probability distributions for $\hat{f}$ and $\hat{a}$ are related by the differential $\mathrm{d}f/\mathrm{d}a$, so even if one is unbiased, the differential, unless it is trivial, ensures that the other is not.

*Example Introduction of bias*
Suppose that $\hat{a}$ is an estimator of $a$ which is unbiased and, for the sake of the illustration, has a probability distribution which is symmetric about the true value $a_0$. Then the chance of it being more than 10% too low and of being more than 10% too high are equally likely, however,

if $\hat{a}$ is 10% too high, then $\hat{a} = 1.1a_0$, and $\hat{a}^2 = 1.21a_0^2$
and if $\hat{a}$ is 10% too low, then $\hat{a} = 0.9a_0$, and $\hat{a}^2 = 0.81a_0^2$.

Thus, for the variable $\hat{a}^2$ the chances of being more than 21% too high and more than 19% too low are equal—the probability distribution in $\hat{a}^2$ is asymmetric and biased.

### ★5.3.2 Maximum Likelihood at Large $N$

As $N \to \infty$, any consistent estimator is unbiased, so the bias of a consistent ML estimator vanishes in this large $N$ or *asymptotic limit*. It will now be shown that it is also efficient.

Suppose that the true value of $a$ is $a_0$. The estimator $\hat{a}$ satisfies

$$\left.\frac{\mathrm{d}\ln L}{\mathrm{d}a}\right|_{a=\hat{a}} = 0$$

which can be expressed as a Taylor expansion about $a = a_0$. Consistency ensures that $\hat{a} - a_0$ becomes small for large $N$, so only the first two terms need be kept:

$$\left.\frac{\mathrm{d}\ln L}{\mathrm{d}a}\right|_{a_0} + (\hat{a} - a_0)\left.\frac{\mathrm{d}^2\ln L}{\mathrm{d}a^2}\right|_{a_0} = 0. \tag{5.27}$$

$\hat{a}$ thus differs from the true value $a_0$ because the derivative of $\ln L$ at $a_0$ differs from zero due to statistical fluctuations in the data.

Now, by equation 5.6, $(d\ln L)/da$ has expectation value zero (the expectation value being evaluated with the same value of $a$ as that for which the differential is evaluated). Furthermore, as it is obtained by summing the $N$ independent terms for each of the $x_i$, the CLT (section 4.1) guarantees that its distribution will be Gaussian. The variance is

$$V\left\langle\left.\frac{\mathrm{d}\ln L}{\mathrm{d}a}\right|_{a_0}\right\rangle = \left\langle\left(\frac{\mathrm{d}\ln L}{\mathrm{d}a}\right)^2\right\rangle = -\left\langle\frac{\mathrm{d}^2\ln L}{\mathrm{d}a^2}\right\rangle$$

using equation 5.8. The expectation values are performed at $a = a_0$.

As $(\hat{a} - a_0)$ is proportional to $[(\mathrm{d}\ln L)/\mathrm{d}a]|_{a_0}$, by equation 5.27, it is also described by a Gaussian distribution with mean zero and variance

$$-\left\langle\frac{\mathrm{d}^2\ln L}{\mathrm{d}a^2}\right\rangle \bigg/ \left(\left.\frac{\mathrm{d}^2\ln L}{\mathrm{d}a^2}\right|_{a_0}\right)^2.$$

Now, in the large $N$ limit, expectation values converge to true values, so the expectation value in the numerator is replaceable by its value at $a_0$. This cancels one of the denominator factors, giving for the variance of $\hat{a}$

$$\sigma_{\hat{a}}^2 = V(\hat{a}) = -\frac{1}{\left(\left.\frac{\mathrm{d}^2 \ln L}{\mathrm{d}a^2}\right|_{a_0}\right)} = -\frac{1}{\left\langle\frac{\mathrm{d}^2 \ln L}{\mathrm{d}a^2}\right\rangle} \tag{5.28}$$

which is just the minimum variance bound.

### ★5.3.3 Errors on the ML Estimators

In the previous section it is shown that for large samples the variance of the ML estimator is equal to the MVB. This means that the Schwarz inequality (equation 5.7) must be saturated (i.e. the inequality becomes an equality), which happens only if the quantities $u$ and $v$ are directly proportional, for all $x$:

$$v(x; a) = A(a)u(x; a)$$

as then $\lambda = -A$ makes the integral zero. In our case this means

$$\frac{\mathrm{d} \ln L}{\mathrm{d}a} = A(a)(\hat{a} - a). \tag{5.29}$$

Differentiating with respect to $a$ ($\hat{a}$ is independent of $a$) gives

$$\frac{\mathrm{d}A}{\mathrm{d}a}(\hat{a} - a) - A = \frac{\mathrm{d}^2 \ln L}{\mathrm{d}a^2}. \tag{5.30}$$

Taking expectation values ($A$ is independent of $x$ and $\hat{a}$ is unbiased) shows that this proportionality factor is an old friend:

$$A = -\left\langle\frac{\mathrm{d}^2 \ln L}{\mathrm{d}a^2}\right\rangle. \tag{5.31}$$

Furthermore, for the value $a = \hat{a}$, equation 5.30 gives

$$A = -\left.\frac{\mathrm{d}^2 \ln L}{\mathrm{d}a^2}\right|_{a=\hat{a}}$$

so

$$\left\langle\frac{\mathrm{d}^2 \ln L}{\mathrm{d}a^2}\right\rangle = \left.\frac{\mathrm{d}^2 \ln L}{\mathrm{d}a^2}\right|_{a=\hat{a}} \tag{5.32}$$

which is useful in evaluating $V(\hat{a})$, as it applies not only for large $N$ but for any unbiased, efficient ML estimator.

In the previous section it was shown that, thanks to the CLT, the probability distribution for $\hat{a}$ was Gaussian. For this to be exactly true, the second

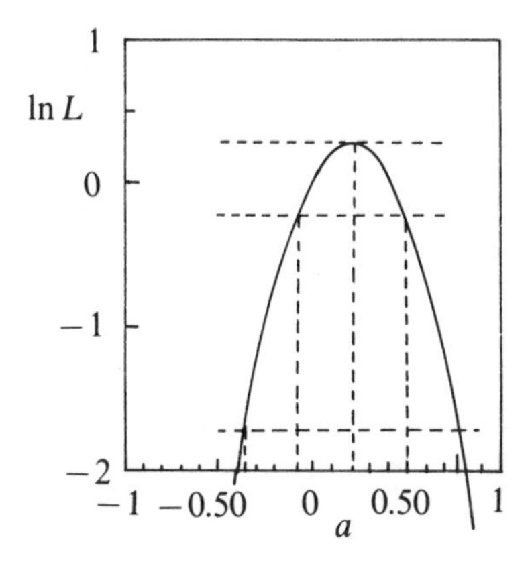


Fig. 5.2. A log likelihood function showing the $1\sigma$ and $2\sigma$ limits.

derivative $(\mathrm{d}^2 \ln L)/\mathrm{d}a^2$ has to be a constant; for it to be a good approximation it has to change very little over the relevant range of $a$ close to $a_0$.

To this approximation, therefore, the quantity $A$ is constant, and equation 5.29 can be integrated and exponentiated to give

$$L(x_1, \ldots, x_N; a) \propto \exp A[a - \hat{a}(x_1, \ldots, x_n)]^2/2 \tag{5.33}$$

showing that in the large $N$ limit the likelihood function you obtain from your data is a Gaussian, the log likelihood is a parabola.

The standard deviation of this Gaussian is just $1/\sqrt{A}$. This is also (cf. equations 5.28 and 5.31) the standard deviation of the estimator $\hat{a}$. You can thus read off the errors on your estimate from the graph of the log likelihood you obtained from the data. At the point $1\sigma$ from the peak the log likelihood has fallen by 0.5 from its maximum. At the $2\sigma$ points it has fallen by 2, at $3\sigma$ by 4.5 and so on.

If the large $N$ limit has not been reached, then the likelihood function is not Gaussian and the log likelihood is not a parabola. Such cases can be handled using the invariance property. Although $\ln L(a)$ may be non-parabolic, there is presumably some alternative parameter $a'$ which transforms the shape to a parabola. If one were to work with such a parameter one would quite happily find the values of $a'$ at which $\ln L$ is 0.5 below its maximum, and quote them as $\pm\sigma$ limits. You could then find the corresponding values of $a$. These, by invariance, are the values of $a$ for which $\ln L$ falls by 0.5 from its peak. We can now forget all about $a'$—the one standard deviation limits are those for which $\ln L$ falls by 0.5 from its peak, for finite $N$ as well as for $N$ large. The important difference between the two cases is that these values may not be symmetric about the peak value $\hat{a}$, giving asymmetric errors, written in the obvious form

$$a = 0.21^{+0.29}_{-0.27}.$$

The $2\sigma$ limits can similarly be read off from the points where the log likelihood falls by 2.0 from its maximum value. Be warned that, for finite $N$, these will not be twice the one standard deviation limits.

### ★ 5.3.4 Several Variables

If you want to estimate the best values for a whole set of parameters (call them $a_1, a_2, \ldots, a_n$) the whole process just generalises. You now have to choose the values of all the variables such that $L$ is maximum, which is usually done by solving the $n$ simultaneous equations

$$\frac{\partial \ln L(x_1, x_2, \ldots, x_N;\, a_1, a_2, \ldots, a_n)}{\partial a_j} = 0 \qquad \text{for all } j = 1, \ldots, n. \quad (5.34)$$

The variance generalises in the obvious way. In the large $N$ limit the likelihood is a multidimensional Gaussian (see section 3.4.6). Just as for a single variable the variance is minus the inverse of the expectation value of the second derivative (equation 5.28), so for many variables the covariance matrix is minus the inverse of the matrix of second derivatives, as one would expect from equation 3.22:

$$\operatorname{cov}^{-1}(a_i, a_j) = -\left\langle \frac{\partial^2 \ln L}{\partial a_i \,\partial a_j} \right\rangle = \left\langle \frac{\partial \ln L}{\partial a_i} \frac{\partial \ln L}{\partial a_j} \right\rangle = -\left. \frac{\partial^2 \ln L}{\partial a_i \,\partial a_j} \right|_{a=\hat{a}} \quad (5.35)$$

and equivalent forms using $P$ (cf. equation 5.9). Note that this is a *matrix* inversion.

The likelihood function is harder to plot, though for two variables it can be shown using contours. For large $N$ the function is a two-dimensional Gaussian or binormal (see section 3.4.7). For small $N$ it can take various shapes, and is prone to the problem of having more than one local maximum. As in the one-dimensional case, one determines errors by finding the points where $\ln L$ is 0.5 below its peak value—these now form a closed curve or, in a really bad case, curves. For large $N$ this is an ellipse, for which, with reference to section 3.4.7, the $1\sigma$ limits can be read off as the limits of the ellipse on the appropriate axis. This is further discussed in section 7.2.7.

*Example Variance of the mean and $\sigma$ estimates for a Gaussian*

$$\ln L = -\sum \frac{(x_i - \mu)^2}{2\sigma^2} - N \ln \sigma - \frac{N}{2} \ln 2\pi.$$

To find the variance we need the second differential, evaluated at the maximum, where

$\mu = \bar{x}$, $\sigma^2 = \overline{x^2} - \mu^2$. This gives

$$\left\langle \frac{\partial^2 L}{\partial \mu^2} \right\rangle = -\frac{N}{\sigma^2} \qquad \left\langle \frac{\partial^2 L}{\partial \sigma^2} \right\rangle = -\frac{2N}{\sigma^2} \qquad \left\langle \frac{\partial^2 L}{\partial \mu \, \partial \sigma} \right\rangle = -\frac{2\sum \langle x_i - \mu \rangle}{\sigma^3} = 0.$$

As the matrix is diagonal, the inversion is trivial, the covariance between $\mu$ and $\sigma$ is zero, and the variances are

$$V(\mu) = -\left\langle \frac{\partial^2 L}{\partial \mu^2} \right\rangle^{-1} = \frac{\sigma^2}{N} \qquad V(\sigma) = -\left\langle \frac{\partial^2 L}{\partial \sigma^2} \right\rangle^{-1} = \frac{\sigma^2}{2N}.$$

### ★ 5.3.5 Notes on Maximum Likelihood

The principle of maximum likelihood is not a rule that requires justification —it does not need to be proved. It is merely a sensible way of producing an estimator. But although the name 'maximum likelihood' has a nice ring to it—it suggest that your estimate is the 'most likely' value—this is an unfair interpretation; it is the estimate that makes your data most likely—another thing altogether.

For large samples, $\hat{a}$ has a probability distribution that is unbiased and normally distributed about the true value $a$, with variance equal to the minimum variance bound. You cannot ask for better than that. So in the large $N$ limit (sometimes referred to as the 'asymptotic limit') ML can indeed claim to be the 'best' estimator, though for smaller samples this is not necessarily true.

It has other obvious advantages. There is no loss of information through binning; all the experimental information is used. It is very suitable for problems where several variables are to be estimated. The method readily gives the errors on its estimate: the $1\sigma$ points are those where the log likelihood falls by 0.5 from its peak value.

Even so, it is not all roses. For small $N$, ML estimators are (usually) biased. You have been warned. And you have to be careful not to apply large $N$ formulae to small $N$ situations.

You have to know the form of the parent distribution function. If your assumptions about $P(x; a)$ are wrong then there is no way of telling this from the results of the fit; it does not give you any quality factor or goodness-of-fit number.

Sometimes the differentiation of $\ln L$ gives you an equation you can solve analytically. (When differentiating the likelihood, do not forget the normalisation factors.) Sometimes you cannot, and it has to be done numerically and/or graphically.

This probably requires you to use a computer and code your likelihood function as a program. If your computer complains about non-existent logarithms of negative numbers, your form for the likelihood function has gone negative and is unphysical, so think again.

For heavy calculations, computer minimisation techniques exist. Do not be tempted to write your own (particularly if you have more than one variable). Use a good software library, e.g. that supplied by the Numerical Algorithms Group. However, for some reason, the supplied programs usually *minimise* functions and you want to *maximise* the likelihood, so remember to put a minus sign in front!

## ★5.4 EXTENDED MAXIMUM LIKELIHOOD

In the standard method of maximum likelihood (ML) the probability or probability density of $x$, $P(x; a)$, is normalised to 1:

$$\int P(x; a)\, \mathrm{d}x = 1. \tag{5.36}$$

In the method of *extended maximum likelihood* (EML), first proposed by Fermi, this requirement is relaxed. Instead of the function $P(x; a)$ one uses a function $Q(x; a)$ for which the normalisation is not fixed. Increasing or decreasing $Q(x; a)$ in some region of $x$ increases or decreases the probability of events occurring in that region, and thus increases or decreases the total expected number of events, $\nu$, which is given by the total integral of $Q(x)$:

$$\int Q(x; a)\, \mathrm{d}x = \nu. \tag{5.37}$$

This is appropriate in experiments where the number of events is indeed unknown *a priori*, and in such cases EML is an improvement over ML. Such experiments are typically those in which data are taken for a certain time, during which events occur 'at random' in some way, as opposed to those which take a prearranged number of events (see problem 3.5). In such cases the incorporation of the fact that the number observed has the actual value $N$ improves the estimates of the parameters. At the same time, the resulting value of $\nu$, which incorporates information from the shape, is a better estimate of the 'true' total number than the basic number of observed events $N$.

The information can be incorporated by combining the standard maximum likelihood with the knowledge that a particular $Q(x; a)$ predicts $\nu$ events in the observed range, and accordingly multiplies the likelihood of a given data sample of $N$ events by the Poisson probability of obtaining $N$ events from a mean of $\nu$:

$$\mathrm{e}^{-\nu} \frac{\nu^N}{N!}.$$

The log likelihood is thus (dropping the $n!$ term as it does not depend on $a$)

$$\begin{aligned}\ln \mathscr{L} &= \sum \ln P(x_i; a) - \nu + N \ln \nu \\ &= \sum \ln [\nu P(x_i; a)] - \nu. \\ &= \sum \ln Q(x_i; a) - \nu. \end{aligned} \tag{5.38}$$

You can see that increasing the normalisation of $Q$ will increase the likelihood by increasing the sum—it makes the observed events more likely—but at the same time it decreases it through the $\nu$ term—it makes it less likely that no other events would have been seen. The maximisation finds the appropriate balance between the two effects.

Thus it differs from the standard likelihood $L$ in two ways, firstly the inclusion of $\nu$ and secondly the fact that $Q(x; a)$ is normalised according to $\nu$ (equation 5.37) rather than 1 (equation 5.36).

The EML and ML methods are so similar that it is plausible that their properties are generally the same. EML estimators also have a bias for finite $N$ which vanishes in the large $N$ limit (unless they are inconsistent, which is most unusual). They are asymptotically efficient, and the error on an estimate from the data is given by the square root of minus the second derivative of the actual likelihood, as described in sections 5.3.1, 5.3.2, and 5.3.3. The expectation values for the elements of the inverse covariance matrix in terms of $Q(x; a)$ are

$$\left\langle \frac{\partial \ln \mathscr{L}}{\partial a_i} \frac{\partial \ln \mathscr{L}}{\partial a_j} \right\rangle = \int \frac{\partial \ln Q}{\partial a_i} \frac{\partial \ln Q}{\partial a_j} Q(x; a)\,\mathrm{d}x. \tag{5.39}$$

These are the same as 5.35, apart from a factor of $N$ due to the normalisation difference.

An alternative application of EML is to problems where the number of events is fixed, but the algebra of equation 5.36 is too complicated for the normalisation to be automatically incorporated in the parametrisation. Such problems can be handled by the EML method, relaxing the overall normalisation condition and letting it vary by including an extra parameter. This can be a simple overall multiplication factor, but if there is any way of altering the parameters such that the normalisation changes but the shape does not, then this is equivalent. Under these conditions it can be shown that the ML and EML solutions will have the same solution for a maximum, and the fitted number of events $\nu$ will be the true number $N$. However, the EML errors are larger than the ML errors, because the problem believes the numbers are capable of fluctuation, and this uncertainty feeds into the parameters. For a genuine EML problem this is indeed the case, but when EML is being used merely to avoid problems with normalisation the ML errors are appropriate.

## ★5.5 THE METHOD OF MOMENTS

The *method of moments* is another method of estimation. Given a distribution function $P(x; a)$, where $a$ is to be estimated, the expectation value of the mean

$$\langle x \rangle = \int x P(x; a) \, dx$$

is a known function of $a$. One can then estimate $a$ by making the expected mean equal to the actual observed mean obtained from the data:

$$\langle x \rangle = \int x P(x; \hat{a}) \, dx = \bar{x}.$$

If there are $n$ unknown parameters $a_1, a_2, \ldots, a_n$ then one solves the $n$ simultaneous equations

$$\langle x^r \rangle = \int x^r P(x; \hat{a}_1 \hat{a}_2, \ldots, \hat{a}_n) \, dx = \overline{x^r} \qquad \text{for } r = 1, \ldots, n.$$

The normalisation condition

$$\int P(x; \hat{a}_1, \hat{a}_2, \ldots, \hat{a}_n) \, dx = 1$$

can be regarded as the equation for the zeroth moment.

In some cases, for example when $P(x; a_1, a_2, \ldots, a_n)$ is a polynomial in $x$, these equations are simple and can be solved directly using matrices, in contrast to the messy numerical iteration of a maximum likelihood fit. The errors on the estimates can be found from the errors on the moments (equations 5.15 and 5.16).

There are obvious drawbacks. Only a few numbers are used, not the whole data, so one is obviously losing information. The variances of the higher moments are large, inducing large errors on the estimates of the parameters.

In a given problem it may be possible to improve the method by choosing a set of functions $\{g_1(x), g_2(x), \ldots, g_n(x)\}$ and using their moments $\overline{g_r}$ and $\langle g_r \rangle$ rather than the $x^r$ in the method.

In particular, you may be able to write $P(x)$ in the form

$$P(x) = 1 + a_1 h_1(x) + a_2 h_2(x) + \cdots + a_n h_n(x)$$

where the $h_i(x)$ are orthonormal, and of zero integral:

$$\int h_i(x) h_j(x) \, dx = \delta_{ij} \qquad \int h_i(x) \, dx = 0.$$

The solution to this problem is then immediate: the integration properties give $\langle h_i \rangle = a_i$, so the estimator is $\widehat{a_i} = \overline{h_i}$, and these are unbiased.

## ★5.6 MAXIMUM LIKELIHOOD AND LEAST SQUARES

Suppose a data sample consists of a set of $(x, y)$ pairs, $\{(x_i, y_i)\}$. The $x_i$ are known exactly, whereas the $y_i$ have been measured, each with some known resolution $\sigma_i$—this is quite a common situation in practice. Suppose further that $y$ is believed to be given by some function $f$ of $x$, which also depends on a parameter (or, in general, several parameters) $a$, and that $a$ is what we want to estimate. This is the ideal $y$, however, and our actual measurements have been smeared by the resolution.

Invoking the central limit theorem to show that the distributions of the measured $y$ values about their ideals is Gaussian, the probability of a particular $y_i$, for a given $x_i$, is

$$P(y_i; a) = \frac{1}{\sigma_i\sqrt{2\pi}} e^{-[y_i - f(x_i;a)]^2/2\sigma_i^2}.$$

The logarithm of the likelihood for the complete data set is thus

$$\ln L = -\frac{1}{2}\sum\left[\frac{y_i - f(x_i)}{\sigma_i}\right]^2 - \sum \ln \sigma_i\sqrt{2\pi}.$$

To *maximise* the likelihood one has to *minimise* the quantity

$$\sum_i\left[\frac{y_i - f(x_i; a)}{\sigma_i}\right]^2$$

i.e. to make the weighted sum of the squared differences as small as possible—hence this is known as the *method of least squares*.

This can be regarded as a 'proof' of the principle of least squares, to be treated in the next chapter. Others prefer to regard it as an assumption in its own right, needing no further justification. For example, G.U. Yule and M.G. Kendall† put this viewpoint very strongly:

> It was formerly the custom, and is still so in works on the theory of observations, to derive the method of least squares from certain theoretical considerations, the assumed normality of the errors of observations being one such. It is, however, more than doubtful whether the conditions for the theoretical validity of the method are realised in statistical practice, and the student would do well to regard the method as recommended chiefly by its comparative simplicity and by the fact that it has stood the test of experience.

## ★5.7 STRATIFIED SAMPLING—BEATING $\sqrt{N}$

Suppose you are trying to find the mean of some quantity for a population on the basis of a relatively small sample—for example, the average height

†*An Introduction to the Theory of Statistics*, 14th edition, 1958, p. 343.

of all the students currently attending a particular university. You set out and take $N$ measurements, and find your sample has some mean $\mu$ and some estimated standard deviation $s$ and can quote a result $\mu \pm s/\sqrt{N}$.

But you can do better than that.

Students come in two varieties—male and female. It is an observed and undisputed fact that the average heights of male and female students are different. Now, you can easily establish the relative proportions of male and female students by consulting the appropriate records. (The current Manchester male: female ratio is 64 to 36%.) The ratio in your sample fluctuates about this quantity, and such fluctuations will add to the fluctuations in the average height: you can avoid them by measuring a sample which contains a specified male: female ratio, and thus reduce the error on your result. This is called *Stratified Sampling.*

To quantify this, first look at what happens if you just sample randomly. Suppose that the distribution is a mixture of $P_1(x)$ and $P_2(x)$, with means $\mu_1$ and $\mu_2$, in the proportions $f_1$ and $f_2$ (so $f_1 + f_2 = 1$) and thus total mean $\mu = f_1\mu_1 + f_2\mu_2$. If you take $N$ measurements then the expected variance of the total is

$$V(x) = \int [f_1 P_1(x) + f_2 P_2(x)](x - \mu)^2 \, dx.$$

Thus is just $f_1$ times the expected value of $(x - \mu)^2$ for type 1 plus the same for type 2, leading to

$$V(x) = f_1 V_1(x) + f_2 V_2(x) + f_1 f_2 (\mu_1 - \mu_2)^2.$$

On the other hand, if you use stratified sampling, in the ratio $f_1$ to $f_2$, then the CLT gives the variance as just the first two terms, without the third. The third term is there because of the possible (binomial) fluctuations in taking the sample. Stratified sampling eliminates it, and beats $\sqrt{N}$.

In general, if you know that the population is divisible into two (or more) types, of which you know the overall proportions, and have reason to believe that these types have different mean values, then you can win by sampling a fixed number from each type. This cuts down the variation in your sample which is due to the different possible numbers of each type you happen to pick.

Note that this works provided you know the numbers $f_1$ and $f_2$ from some other information, and that the magnitude of $\mu_1 - \mu_2$ is significant. You gain nothing but trouble and wasted time if the two samples are not significantly different.

What is the best stratification? Is it sensible to sample in the same ratio as the types naturally occur? Suppose you take $m_1$ measurements of type 1 and $m_2$ of type 2. You then estimate $\mu_1$ and $\mu_2$ with variances $V_1/m_1$ and $V_2/m_2$. The estimates are combined to give

$$\hat{\mu} = f_1\hat{\mu}_1 + f_2\hat{\mu}.$$

The variance of this is just

$$V = \frac{f_1^2 V_1}{m_1} + \frac{f_2^2 V_2}{m_2}.$$

We want to choose $m_1$ and $m_2$ to minimise this, subject to the fact that $m_1 + m_2$ is constrained by the budget to be some fixed number $N$. Putting $m_2 = N - m_1$ and differentiating the variance with respect to $m_1$ leads to

$$\frac{m_1}{m_2} = \frac{f_1\sqrt{V_1}}{f_2\sqrt{V_2}} = \frac{f_1\sigma_1}{f_2\sigma_2}.$$

If the $\sigma$ are the same (or if you have no knowledge about whether they may be different) then this confirms that the ratio in your sample is best taken as the ratio in the data. If you have trustworthy reasons to believe that one has a broader distribution that the other, then it is good sense to take more samples of the broader type.

If there are several types, then the variance is given by

$$V = f_1^2 \frac{V_1}{m_1} + f_2^2 \frac{V_2}{m_2} + f_3^2 \frac{V_3}{m_3} + \cdots.$$

This can be minimised using the method of Lagrangian multipliers. Given the constraint that $m_1 + m_2 + m_3 + \cdots = N = constant$, so that

$$\frac{\partial N}{\partial m_i} = 1 \quad \forall i$$

then this gives at once

$$m_i \propto f_i \sigma_i$$

as one would expect.

In practice one might settle for a less than optimal sampling setup. Even random sampling will give a better than $\sqrt{N}$ result if you note down the type at the same time as taking the measurement.

## 5.8 PROBLEMS

*5.1*
A binomial experiment results in $N_S$ successes and $N_F$ failures. Show that

$$\hat{p} = \frac{N_S}{N_S + N_F}$$

is a consistent, unbiased estimator for the individual probability $p$.

*5.2*
Given the seven numbers

20.0 19.7 20.6 18.5 21.2 20.8 20.7

(a) Calculate the mean. What is the error on the mean, if the numbers represent (Gaussian) measurements of a single quantity, with a resolution of 0.8?

(b) Estimate the standard deviation, given that the true mean is 20.0. What is the error on the estimate?

(c) Estimate the standard deviation given no prior knowledge of the true mean. What is the error on the estimate?

(d) Find the error on the mean given no prior knowledge of the standard deviation.

★ *5.3*

Show that an unbiased estimator of $\langle (x-\mu)^3 \rangle$ is

$$\frac{N}{(N-1)(N-2)} \sum (x_i - \bar{x})^3.$$

★ *5.4*

Show that the error on the variance (equation 5.17) can be written in terms of the curtosis (section 2.5.2) as

$$V(\widehat{V(x)}) = \frac{\sigma^4(c+2)}{N}.$$

★ *5.5*

If you have two independent measurements of equal accuracy, one of $\sin\theta$ and one of $\cos\theta$, find the ML estimate for $\theta$.

★ *5.6*

Show that if the lifetime $\tau$ in exponential decay (with complete acceptance) is found by maximum likelihood, the variance for large samples is

$$V(\hat{\tau}) = \frac{\tau^2}{N}.$$

CHAPTER 6

# Least Squares

The method of least squares is a way of determining unknown parameters from a set of data. It really belongs in the previous chapter with other methods of estimation, but it is so useful and important that it gets a chapter of its own. In its most basic form, it is used when you have two variables $x$ and $y$, and

(a) a set of precisely known $x$ values,
(b) a corresponding set of $y$ values, measured with some accuracy $\sigma$,
(c) a function $f(x; a)$ which predicts the value of $y$ for any $x$; the form of this function is known but it involves an unknown parameter $a$, which you are trying to determine.

## 6.1 OUTLINE OF THE METHOD

The *principle of least squares* can be derived from the principle of maximum likelihood (if the measurements follow a Gaussian distribution: see section 5.6). Alternatively, it can just be regarded as an obviously sensible estimator. 'Least squares' means just what it says: you minimise the (suitably weighted) squared difference between a set of measurements and their predicted values. This is done by varying the parameters you want to estimate: the predicted values are adjusted so as to be close to the measurements; squaring the differences means that greater importance is placed on removing the large deviations.

This provides a means of estimating a parameter $a$ in a function $f(x;a)$ which predicts the true value of $y$ for any $x$. The data it uses to do this are a set of $N$ precise values of $x - \{x_1, x_2, \ldots, x_N\}$—with a corresponding set of measurements of $y$—$\{y_1, y_2, \ldots, y_N\}$ where each value $y_i$ has been measured with some accuracy $\sigma_i$. You take the sum over all points of the squared difference between the measurement $y_i$ and the prediction $f(x_i;a)$, scaled by the expected error $\sigma_i$. This sum is called $\chi^2$:

$$\chi^2 = \sum_{i=1}^{N} \left[ \frac{y_i - f(x_i;a)}{\sigma_i} \right]^2. \tag{6.1}$$

Then choose the value of $a$ which gives the smallest $\chi^2$. If all the $\sigma_i$ are the same, then $\sigma$ comes out as a common factor and a lot of the algebra is much simpler. If the derivatives of $f$ with respect to $a$ are known, as they usually are, then the minimisation problem becomes one of finding a solution of the equation

$$\frac{d\chi^2}{da} = 0 \tag{6.2}$$

that is

$$\sum_i \frac{1}{\sigma_i^2} \frac{df(x_i;a)}{da} [y_i - f(x_i;a)] = 0. \tag{6.3}$$

The resulting estimate of $a$—denote it $\hat{a}$—will, one hopes, be close to the true value, but it is unlikely to coincide exactly. The expected error on the estimate is known, as the solution of equation 6.3 gives $\hat{a}$ as a function of all the $y_i$. These have known errors $\sigma_i$, and the formula for combination of errors (section 4.3) gives the resulting error on $\hat{a}$.

If the function contains not merely one unknown parameter $a$ but several—$a_1, a_2, \ldots, a_n$—then there are $n$ simultaneous equations in the $n$ unknowns which are solved to give the estimates.

### 6.1.1 Fitting $y = mx$—simple proportion

To show how least squares works, consider the case where $y$ is proportional to $x$, i.e. $f(x) = mx$. This is a simple example of a fit to one unknown parameter $m$, and also a case commonly met with in practice. The quantity to be minimized, by adjusting $m$, is

$$\chi^2 = \sum_i \frac{(y_i - mx_i)^2}{\sigma_i^2}.$$

Differentiating this with respect to $m$ gives

$$\frac{\mathrm{d}\chi^2}{\mathrm{d}m} = \sum -2x_i \frac{y_i - mx_i}{\sigma_i^2}.$$

If all the $\sigma_i$ are the same, they are a common factor and can be taken outside the sum, giving

$$-\frac{2}{\sigma^2}\sum (x_i y_i - mx_i^2).$$

For the least squares estimate of the slope, $\hat{m}$, this is zero:

$$\sum (x_i y_i - \hat{m}x_i^2) = 0$$
$$\sum x_i y_i = \hat{m}\sum x_i^2.$$

So the least squares estimate of the proportionality constant, dividing by $N$ to turn sums to averages, is

$$\hat{m} = \frac{\overline{xy}}{\overline{x^2}}. \tag{6.4}$$

How accurate is this? Writing the result as

$$\hat{m} = \sum \frac{x_i}{N\overline{x^2}} y_i$$

then combination of the errors from each $y_i$ (see section 4.3.2) gives

$$V(\hat{m}) = \sum \left(\frac{x_i}{N\overline{x^2}}\right)^2 \sigma^2 = \frac{\sigma^2}{N\overline{x^2}}. \tag{6.5}$$

## 6.2 THE STRAIGHT LINE FIT

Fitting a straight line is probably the most common application of least squares fitting, so it deserves treatment in full detail. As shown in Figure 6.1,

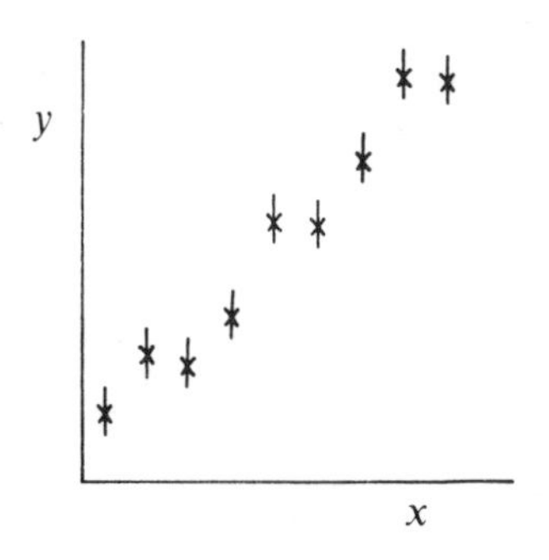


Fig. 6.1. A common problem.

the measurements $y_i$ all have the same error $\sigma$, and are supposed to lie on a straight line $y = mx + c$, where $m$ (the slope) and $c$ (the intercept) are unknown and to be determined. Their least squares estimates are derived in the following sections. They are summarised here for reference.

$$\hat{m} = \frac{\overline{xy} - \bar{x}\bar{y}}{\overline{x^2} - \bar{x}^2}$$

$$\hat{c} = \bar{y} - \hat{m}\bar{x}$$

with errors

$$\sigma_m^2 = V(\hat{m}) = \frac{\sigma^2}{N(\overline{x^2} - \bar{x}^2)}$$

$$\sigma_c^2 = V(\hat{c}) = \frac{\sigma^2 \overline{x^2}}{N(\overline{x^2} - \bar{x}^2)}$$

$$\operatorname{cov}(\hat{m}, \hat{c}) = -\frac{\sigma^2 \bar{x}}{N(\overline{x^2} - \bar{x}^2)}$$

$$\rho_{\hat{m},\hat{c}} = -\frac{\bar{x}}{\sqrt{\overline{x^2}}}$$

and the $\chi^2$ for the best fit is

$$\chi^2 = \frac{V(y)}{\sigma^2}(1 - \rho_{x,y}^2).$$

### 6.2.1 The Slope and Intercept for a Straight Line

Deriving these results is straightforward, and is similar to that followed in detail in section 6.1.1.

The sum to be minimised is

$$\sum_i (y_i - mx_i - c)^2.$$

Differentiating this with respect to $c$ and setting it to zero gives

$$\sum_i -2(y_i - \hat{m}x_i - \hat{c}) = 0$$

which, on dividing by $N$, gives

$$\bar{y} - \hat{m}\bar{x} - \hat{c} = 0.$$

Differentiating the sum with respect to $m$ gives

$$\sum_i -2x_i(y_i - \hat{m}x_i - \hat{c}) = 0$$

and thus

$$\overline{xy} - \hat{m}\overline{x^2} - \hat{c}\bar{x} = 0.$$

Eliminating $\hat{c}$ from the two equations gives the result for the slope

$$\hat{m} = \frac{\overline{xy} - \bar{x}\bar{y}}{\overline{x^2} - \bar{x}^2} \tag{6.6}$$

which can also be written in terms of the covariance and variance:

$$\hat{m} = \frac{\operatorname{cov}(x, y)}{V(x)}. \tag{6.7}$$

Substituting the expression for $\hat{m}$ back in the equations gives

$$\hat{c} = \frac{\overline{x^2}\bar{y} - \bar{x}\overline{xy}}{\overline{x^2} - \bar{x}^2} \tag{6.8}$$

though a form which is usually more convenient is

$$\hat{c} = \bar{y} - \hat{m}\bar{x} \tag{6.9}$$

showing that the line goes through the centre of gravity, $(\bar{x}, \bar{y})$.

### ★ 6.2.2 Derivation of the Errors for a Straight Line

The variance on the estimate of $m$ can be seen by writing equation 6.6 in the form

$$\hat{m} = \sum_i \frac{x_i - \bar{x}}{N(\overline{x^2} - x^2)} y_i$$

whence the law of combination of errors gives

$$V(\hat{m}) = \sum \left[ \frac{x_i - \bar{x}}{N(\overline{x^2} - \bar{x}^2)} \right]^2 \sigma^2.$$

This reduces to

$$V(\hat{m}) = \frac{\sigma^2}{N(\overline{x^2} - \bar{x}^2)}. \tag{6.10}$$

Likewise, applying the law of combination of errors to 6.8 gives

$$V(\hat{c}) = \sum_i \left[ \frac{\overline{x^2} - \bar{x}x_i}{N(\overline{x^2} - \bar{x}^2)} \right]^2 \sigma^2$$
$$= \frac{\sigma^2 \overline{x^2}}{N(\overline{x^2} - \bar{x}^2)}. \tag{6.11}$$

In the same way the covariance is found to be

$$\text{cov}(\hat{m}, \hat{c}) = -\frac{\bar{x}}{N(\overline{x^2} - \bar{x}^2)}\sigma^2. \tag{6.12}$$

Putting 6.6 and 6.9 back in the expression for $\chi^2$ gives the formula

$$\chi^2 = \frac{V(y)}{\sigma^2}(1 - \rho_{x,y}^2). \tag{6.13}$$

This is useful when you want to know the $\chi^2$ that a straight line fit would give through a set of points, without going through the business of finding $\hat{m}$ and $\hat{c}$ and then computing $y_i - \hat{m}x_i - \hat{c}$ at each point.

Please notice that quantities like $V(x)$, $\text{cov}(x, y)$, and $\rho_{x,y}$ are those for the sample as a whole. In particular $V(y) = \overline{y^2} - \bar{y}^2$ describes the spread of the $y$ values of the whole sample, and is not the same as $\sigma^2$, which describes the spread of a single measurement about its true value.

### ★ 6.2.3 Weighted Straight Line Fit

If the $\sigma_i$ are not all equal the sum to be minimised is

$$\sum \frac{(y_i - mx_i - c)^2}{\sigma_i^2}.$$

This gives the same equations for $\hat{m}$ and $\hat{c}$ as before (6.6, 6.8, etc.) except that $\bar{x}$, $\bar{y}$, etc., denote not the simple averages but averages where each point is given a weight $1/\sigma_i^2$, and the normalisation is therefore not the number of points $N$ but the total weight $\sum 1/\sigma_i^2$:

$$\frac{\sum y_i}{N} \to \frac{\sum y_i/\sigma_i^2}{\sum 1/\sigma_i^2}.$$

Also, in the expressions 6.10, 6.11, and 6.12 for the variances, the quantity $\sigma^2$ has to be replaced by

$$\overline{\sigma^2} = \frac{\sum \sigma_i^2/\sigma_i^2}{\sum 1/\sigma_i^2} = \frac{N}{\sum 1/\sigma_i^2}. \tag{6.14}$$

### ★6.2.4 Extrapolation

Having measured the slope and intercept, one often needs to know the extrapolated (or interpolated) value of $y$ at a given $x$—and, of course, the associated error.

Given some $X$, the predicted $Y$ is just $\hat{m}X + \hat{c}$; the error is given by

$$V(Y) = V(\hat{c}) + X^2 V(\hat{m}) + 2X \operatorname{cov}(\hat{m}, \hat{c}). \tag{6.15a}$$

This last term can be very important. If you leave it out you will get too large an error.

Having pointed out the importance of the covariance term, now let us see how to avoid it. If $\bar{x}$ is zero—or if you move to a system where it is—then the covariance term (equation 6.12) vanishes. That means you do not have to remember to include it later. Finding $\bar{x}$ and then fitting

$$y = \hat{m}(x - \bar{x}) + \hat{c}'$$

instead of the more usual form would give uncorrelated estimates for $\hat{m}$ and $\hat{c}'$. The variance of $\hat{m}$ is the same as before, and the error on $\hat{c}'$ is just $\sigma/\sqrt{N}$. The error on the extrapolation is given by this means as

$$V(Y) = \frac{\sigma^2(X - \bar{x})^2}{N(\overline{x^2} - \bar{x}^2)} + \frac{\sigma^2}{N} \tag{6.15b}$$

and this of course is what you get from putting equations 6.10, 6.11, and 6.12 in equation 6.15*a*.

### ★6.2.5 Systematic Errors and a Straight Line Fit

As an exercise in systematic errors, as discussed in section 4.4, consider a straight line fit where all the $y$ values have a random error $\sigma$, and also share a common systematic error $S$. $\hat{m}$ and $\hat{c}$ are given by equations 6.6 and 6.8. The full error formula, equation 4.11, gives

$$V(\hat{m}) = \frac{1}{N^2(\overline{x^2} - \bar{x}^2)^2} \sum_i \sum_j (x_i - \bar{x})(x_j - \bar{x}) \operatorname{cov}(y_i, y_j).$$

From equation 4.20, we have $\operatorname{cov}(y_i, y_j) = \delta_{ij}\sigma^2 + S^2$ so

$$V(\hat{m}) = \frac{1}{N^2(\overline{x^2} - \bar{x}^2)^2} \left[ \sum_i (x_i - \bar{x})^2 \sigma^2 + \sum_i \sum_j (x_i - \bar{x})(x_j - \bar{x}) S^2 \right].$$

The first sum gives equation 6.10. The second gives zero, as $\sum x_i = \bar{x}$. Now for the constant. Equation 4.11 gives

$$V(\hat{c}) = \frac{1}{N^2(\overline{x^2} - \bar{x}^2)^2} \sum_i \sum_j (\overline{x^2} - \bar{x}x_i)(\overline{x^2} - \bar{x}x_j) \operatorname{cov}(y_i, y_j).$$

Again the $\sigma^2$ values along the diagonal give the standard term of equation 6.11, but the other term does not cancel. $\sum(x^2 - \bar{x}x_i)$ is just $N(\overline{x^2} - \bar{x}^2)$, giving an additional term which is just $S^2$.

In summary: a systematic error on $y$ does not affect the slope, but adds (in quadrature) to the uncertainty on the intercept.

Now consider a more complex problem. Suppose you draw a straight line through $(x, y)$ points, which fall into two subsets, one with a systematic error $S_1$ and the other with $S_2$. Measurements from different subsets are uncorrelated.

The error on the slope now contains four terms

$$V(\hat{m}) = \frac{1}{N^2(\overline{x^2} - x^2)^2}\left(\sum_i\sum_j \cdots + \sum_i\sum_{j'} \cdots + \sum_{i'}\sum_j \cdots + \sum_{i'}\sum_{j'} \cdots\right)$$

where the primed indices denote a sum over subset 2 and the unprimed over subset 1. The term of equation 6.11 comes out as usual. There are also two terms from the first and last summations:

$$\frac{1}{N^2(\overline{x^2} - x^2)^2}\sum_i\sum_j (x_i - \bar{x})(x_j - \bar{x})S_1^2 + \sum_{i'}\sum_{j'}(x_{i'} - \bar{x})(x_{j'} - \bar{x})S_2^2$$

which in terms of the two subset average values, $\overline{x_1}$ and $\overline{x_2}$, gives the extra error on the slope due to the two different systematic errors:

$$\frac{N_1^2 S_1^2(\overline{x_1} - \bar{x})^2}{N^2(\overline{x^2} - \bar{x}^2)^2} + \frac{N_2^2 S_2^2(\overline{x_2} - \bar{x})^2}{N^2(\overline{x^2} - \bar{x}^2)^2}.$$

### ★6.2.6 Regression

The data on the left in Figure 6.2 are measurements of the temperature and pressure of a gas at constant volume. The line through the points is the

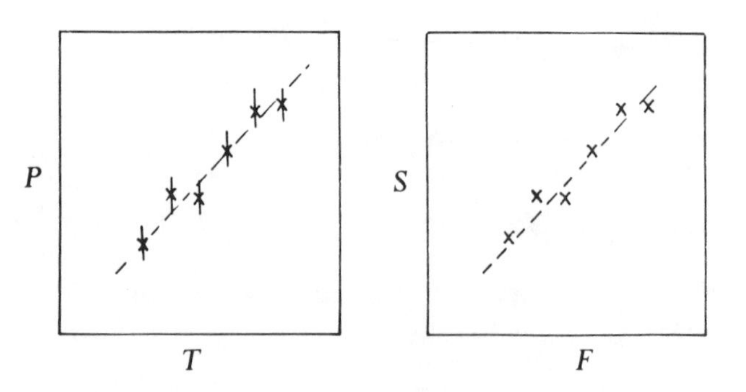


Fig. 6.2. Two sets of data and two straight lines.

'straight line fit'. The data on the right are measurements of the heights of some fathers and their (adult) sons: the line through the points is the 'line of regression'. What is the difference? Mathematically there is none—both lines were evaluated by equations 6.6 to 6.9. However, there is a profound difference in their meaning.

Notice how the left-hand plot has error bars on its measurements and the right-hand one has not. The real difference between the two plots lies here. If $P$ had been measured with greater precision, using a more accurate barometer, then the measured points would move closer to their true relationship. (For an ideal gas this would be a straight line, and for a real gas should be reasonably close to a straight line). If the heights were remeasured using, say, a laser interferometer instead of an ordinary ruler, then there would be no visible effect on the measured points shown.

The true values of pressure and temperature are believed to be related by an exact law, and the line drawn through those points is an estimate of that law. The heights of father and son are related by a trend—if you know that John Smith, Snr, is 6′6″ tall, then you might expect John Smith, Jnr, to be on the tall side, but you could not predict his height very accurately. Thus, although there are similarities, regression is a part of descriptive statistics, like correlation, whereas straight line fitting is a form of estimation.

The name 'regression', which is a bit misleading, came from Galton's original work on just this problem. He found that tall fathers did tend to have tall sons, but that as the correlation is not perfect, a tall father will tend to have a son shorter than himself. He called this 'regression towards the norm', and the name stuck.

## 6.3 FITTING BINNED DATA

The method of least squares is probably the most commonly used method of fitting parameters. Maximum likelihood, as described in Chapter 5, becomes very cumbersome for a large sample as you have to evaluate logarithms of complicated functions for every data element, and least squares provides a more convenient alternative if the data can be binned (see section 2.2) without significant loss of information. Suppose there are $N$ events, and the probability function is $P(x; a)$. They are sorted into bins, numbered from 1 to $N_B$. Bin number $j$ is centered on the point $x_j$, has width $W_j$, and contains $n_j$ events. (The $W_j$ are all the same if the bins are uniform.) The ideal number of events expected in bin $j$ is $f_j = NW_jP(x_j; a)$. The actual number will be described by Poisson statistics, so the squared error is equal to the mean, and the total $\chi^2$, summed over all bins, is

$$\chi^2 = \sum_j \frac{(n_j - f_j)^2}{f_j}. \tag{6.16}$$

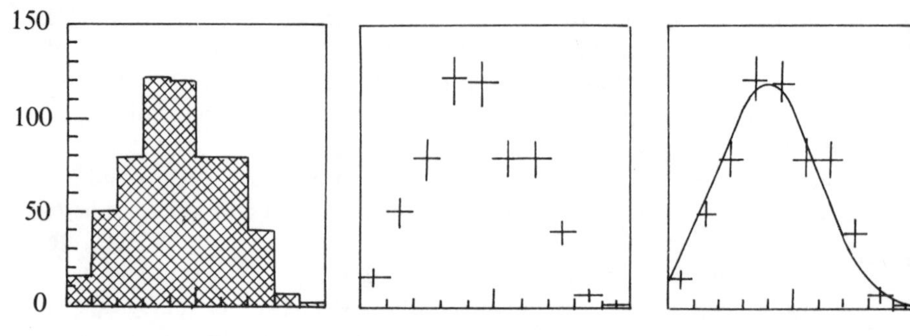


Fig. 6.3. Fitting a histogram.

Some people actually define $\chi^2$ by this formula, but it is really restricted to frequency distributions.

A further shortcut is sometimes applied: one can approximate

$$\chi^2 \approx \sum_j \frac{(n_j - f_j)^2}{n_j} \tag{6.17}$$

which is much easier to handle numerically, but remember that it is valid only if the $n_j$ and $f_j$ are large, and not too different. Do not forget to include the bin width $W$ when working out $f_j$!

Binning can also be used for distributions in two or more dimensions, but it is harder to ensure a reasonable number of points in each bin.

## 6.4 THE $\chi^2$ DISTRIBUTION

The quantity $\chi^2$† is the squared difference between the observed values and their theoretical predictions, suitably weighted by the errors of measurement:

$$\chi^2 = \sum_i \frac{[y_i - f(x_i)]^2}{\sigma_i^2} = \sum_{i=1}^{N} \left( \frac{y_i^{\text{actual}} - y_i^{\text{ideal}}}{\text{expected error}} \right)^2.$$

If the function agrees well with the actual values then $\chi^2$ will be small—so if $\chi^2$ is large at the end of your minimising, it is telling you that there is something wrong with the answer. However, a very small value of $\chi^2$ is also unlikely; the errors should make the measurements deviate from their ideal values to some extent, and a very low $\chi^2$ probably means that these errors have been overestimated. To be more specific requires the distribution for

† $\chi^2$ is generally pronounced 'keye squared', owing to the inability of the English to manage the guttural 'ch'. Those of you with a knowledge of Scots, Welsh, German, or, of course, ancient Greek will have no difficulty pronouncing it correctly, but may confuse your colleagues by doing so.

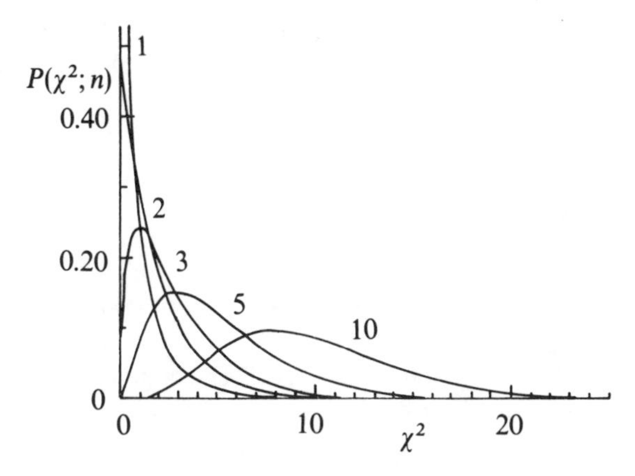


Fig. 6.4. Some $\chi^2$ distributions.

$\chi^2$, which is ($\Gamma(x)$ is the standard gamma function)

$$P(\chi^2; n) = \frac{2^{-n/2}}{\Gamma(n/2)} \chi^{n-2} e^{-\chi^2/2}. \tag{6.18}$$

The distribution depends on $n$, which is the number of points in the sum, $N$, minus the number of variables that have been adjusted to minimise $\chi^2$. It is called the *number of degrees of freedom*. Distributions are shown in Figure 6.4 for $n = 1, 2, 3, 5$, and 10.

The $\chi^2$ distribution has mean $n$ and variance $2n$. Thus one expects a $\chi^2$ per degree of freedom of roughly one, and there are grounds for suspecting that something is wrong if it is much bigger. However, although the central limit theorem ensures that the $\chi^2$ distribution tends to a Gaussian for large $n$, the values of $n$ required for this are extremely large. It is better to use $\sqrt{2\chi^2}$, which has a distribution that is a reasonable approximation to a Gaussian with mean $\sqrt{2n-1}$ and unit variance for values of $n$ greater than about 30. Application to smaller values is discussed in section 8.3.1.

*Example Goodness of fit and* $\chi^2$
A frequency distribution contains a large peak, covering 45 bins. Fitting this to a Gaussian distribution gives a $\chi^2$ of 73. As there are three free parameters (the mean, width, and normalisation of the Gaussian) the number of degrees of freedom is 42. $\sqrt{2n-1}$ has the value 9.1. $\sqrt{2\chi^2}$ is 12.1, larger by three units, which is also $3\sigma$. This strongly suggests that the peak is not well described by a Gaussian.

This can also be used in reverse. Suppose a set of measurements all have the same accuracy, but you do not know what it is. The least squares procedure can still be applied as $\sigma$ can be taken out of equation 6.3 as a common factor.

The total squared deviation per degree of freedom should then just equal $\sigma^2$, so this gives an estimate of the measurement error. This can be very useful—the price you pay is that $\chi^2$ is no longer available as a consistency check.

### ★6.4.1 Proof of the $\chi^2$ Distribution

If each $y_i$ is distributed about its $f_i$ according to a Gaussian, the scaled deviations $u_i = [y_i - f(x_i)]/\sigma_i$ are distributed according to the unit Gaussian. These can be thought of as a point $(u_1, u_2, \ldots, u_N)$ in an $N$-dimensional space, with probability proportional to $\exp(-\chi^2/2)$. The points of constant probability lie on an $N-1$ dimensional hypersphere of radius $\chi$. The probability that a result will occur between $\chi$ and $\chi + \mathrm{d}\chi$ thus depends on $\mathrm{e}^{-\chi^2/2}$, multiplied by the volume of the region concerned, which is proportional to $\chi^{N-1}\,\mathrm{d}\chi$, so $P(\chi) \propto \chi^{N-1}\mathrm{e}^{-\chi^2/2}$. We actually want $P(\chi^2)$; to convert from one to the other we multiply by the differential, so

$$P(\chi^2) = P(\chi)\frac{\mathrm{d}\chi}{\mathrm{d}\chi^2} \propto \chi^{N-1}\mathrm{e}^{-\chi^2/2}\frac{1}{\chi} = \chi^{N-2}\mathrm{e}^{-\chi^2/2}$$

and the result is proved—the constant of proportionality, which ensures the distribution is normalized, can be found by integration.

The important additivity property follows: if two variables are described by $\chi^2$ distributions with $N_1$ and $N_2$ measurements, the sum of the two is obviously described by a $\chi^2$ distribution with $N_1 + N_2$ measurements.

Now suppose that the $u_i$ are still distributed according to the unit Gaussian, but are also subject to a homogeneous linear constraint, i.e.

$$C_1 u_1 + C_2 u_2 + \cdots + C_N u_N = 0$$

where the $C_i$ are constants. These points lie on an $(N-1)$-dimensional subspace of the original, but their probability density is still proportional to $\exp(-\chi^2/2)$. The above analysis still works and the same result for $P(\chi^2)$ is obtained, except that the power of $\chi$ is reduced by one. If there are two such constraint equations, the dimensionality is reduced by 2, and so on.

Thus if the adjustment of a variable to fit the data imposes a linear homogeneous constraint, this decreases the number of degrees of freedom by one. So when $N$ measurements are taken of a quantity, and deviations taken not from the true mean but from the sample mean, this guarantees that $u_1 + u_2 + u_3 + \cdots + u_N = 0$, and a degree of freedom is lost in just this way. Similarly, in fitting binned data, normalising the predictions to the total number of events guarantees that $\sum n_i - f_i = 0$.

In the more general case, suppose that one or more parameters have been determined by the least squares technique. Then equations of the type of equation 6.3 are satisfied; these have the required form provided that the differentials, $\mathrm{d}f/\mathrm{d}a$, are constant, i.e. that the predictions are *linear functions* of the parameters being estimated. Another requirement is that the errors $\sigma_i$ must *not* depend on the parameter, otherwise extra differentials appear.

If either of these two requirements is not satisfied then the 'degrees of freedom' method is, strictly speaking, invalid. However, it is so convenient that it is frequently applied in such cases. This is acceptable provided the non-linearity is not too drastic and any changes in the error are not too large; this is usually the case, and anyway the effect of the discrepancy vanishes for large $N$.

In the $\chi^2$ defined by equation 6.16 for frequency distributions the error in the denominator obviously does vary. However, as the discrepancy $n_i - f_i$ is given not by a Gaussian but by a Poisson distribution, this '$\chi^2$' obeys the $\chi^2$ distribution only in the large $N$ limit anyway. Any further discrepancies are hardly important.

## ★6.5 ERRORS ON $x$ AND $y$

Suppose there are errors on both the $x$ and $y$ measurements. To start with, we take the simplest such case, with the errors on $x$ and $y$ equal, with the value $\sigma$, and fit a straight line.

We now apply maximum likelihood. A measured point, like B in Figure 6.5, could have arisen from any point A along the ideal line. The probability density for this is

$$P(\mathrm{A} \to \mathrm{B}) = \frac{1}{2\pi\sigma^2}\,\mathrm{e}^{-(x_A - x_B)^2/2\sigma^2}\,\mathrm{e}^{-(y_A - y_B)^2/2\sigma^2}$$

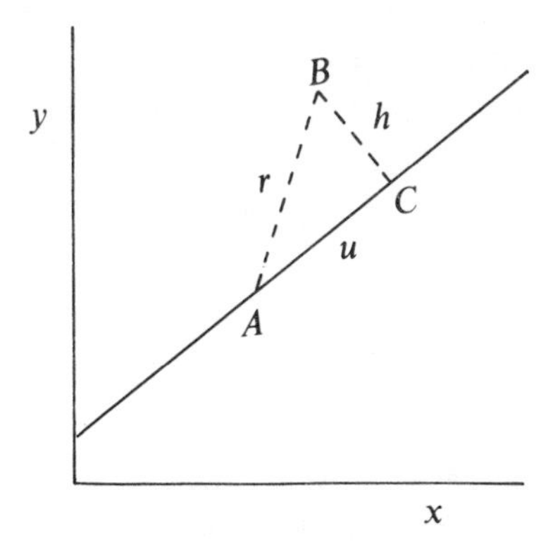


Fig. 6.5. B is a measured point. The solid line shows $y = f(x; a)$, which in this case is a straight line.

which is just

$$\frac{1}{2\pi\sigma^2}e^{-r^2/2\sigma^2}$$

which is also

$$\frac{1}{2\pi\sigma^2}e^{-u^2/2\sigma^2}e^{-h^2/2\sigma^2}.$$

To find the total probability that B could come from any point A, we just integrate $P(A \rightarrow B)$ over all $u$. The integral of $\exp(-u^2/2\sigma^2)$ just gives a constant, $\sigma\sqrt{2\pi}$, so the probability of a point occurring at B is proportional to $\exp(-h^2/2\sigma^2)$. $h$ is the perpendicular distance from the point to the line, i.e. from the measured point B to the point C, which is the point on the line with the greatest probability of producing B. In forming the $\chi^2$ to be minimised one sums the squared distances from each measured point to the point on the line from which it is most likely to have been produced. The fact that it could have been produced by any other point along the line need not be explicitly taken into account.

If the $x$ and $y$ accuracies are not the same, then they can be made so by simple scaling; change to variables $y' = y/\sigma_y$, $x' = x/\sigma_x$, and apply the same logic. Likewise, if the $x$ and $y$ values at a point are correlated, they can be rotated into a frame where they are not. The conclusion persists: one need only consider the point on the line with the greatest probability of producing the data point.

However, this only works for straight lines. The integration over all $u$ only behaves so nicely because the function is a straight line, extending from $-\infty$ to $+\infty$, with a constant probability for points to lie along it. If any of these three conditions is not fulfilled, the 'most likely point' approach is, strictly speaking, invalid. In such cases it may be justifiable to use most likely points, but you have to convince yourself that the effects are small over a few $\sigma$.

Applying this to the straight line with $x$ and $y$ errors equal, the perpendicular distance from a point $x_i$, $y_i$ to the line $y = mx + c$ is

$$h_i = \frac{y - mx_i - c}{\sqrt{1+m^2}}.$$

Thus

$$\chi^2 \propto \sum_i \frac{(y_i - mx_i - c)^2}{1+m^2}.$$

Differentiating with respect to $c$ gives the simple equation $\bar{y} = \hat{m}\bar{x} + \hat{c}$; the line goes through the centre of gravity as usual. Differentiating with respect to $m$ gives a result which can be written as

$$\hat{m} = A \pm \sqrt{A^2 + 1} \tag{6.19}$$

where

$$A = \frac{V(y) - V(x)}{2\,\mathrm{cov}\,(x, y)}.$$

The two solutions for the slope are mutually perpendicular: one gives the best straight line and the other the worst. The plus sign is taken if cov $(x, y)$ is positive, the minus if it is negative. (Note again that $V(y)$, etc., refer to the whole sample and have nothing to do with the measurement error.)

If the measurement errors $\sigma_x$ and $\sigma_y$ are different, then equation 6.19 can be applied to $y/\sigma_y$ and $x/\sigma_x$, as mentioned above, and on transforming back to the original $x$ and $y$ one obtains

$$\hat{m} = \frac{\sigma_y}{\sigma_x}(A \pm \sqrt{A^2 + 1}) \qquad A = \frac{\sigma_x^2 V(y) - \sigma_y^2 V(x)}{2\sigma_x \sigma_y \,\mathrm{cov}\,(x, y)}.$$

If the measurement errors differ from point to point, then there is no analytic solution and numerical methods must be used.

## ★ 6.6 LINEAR LEAST SQUARES AND MATRICES

If there are many unknown quantities to be determined, then matrix notation makes the equations look simple by avoiding summation signs and indices. For the $n$ parameters $a_1, a_2, \ldots, a_n$, one writes simply $\mathbf{a}$, where $\mathbf{a}$ is a vector of $n$ elements. Likewise the $y_i$ are written as $\mathbf{y}$ and the $f(x_i; \mathbf{a})$ as $\mathbf{f}$, where $\mathbf{y}$ and $\mathbf{f}$ are vectors of $N$ elements.

This is also a useful way of working when the $y_i$ are not independent measurements (each with error $\sigma_i$) but are correlated, needing a covariance matrix $\mathbf{V}$ to describe them. For this case, we go back to the principle of maximum likelihood, from which the least squares principle was obtained. This tells us to minimise the exponent in the Gaussian likelihood function, and (see section 3.4.6) this is given by

$$\sum_i \sum_j [y_i - f(x_i; \mathbf{a})] V_{ij}^{-1} [y_j - f(x_j; \mathbf{a})]$$

which can be neatly expressed in matrix notation as

$$\chi^2 = (\tilde{\mathbf{y}} - \tilde{\mathbf{f}})\mathbf{V}^{-1}(\mathbf{y} - \mathbf{f}). \tag{6.20}$$

Independent measurements are a special case of this, where $\mathbf{V}$ is diagonal and $V_{ij} = \sigma_i^2 \delta_{ij}$, $V_{ij}^{-1} = (1/\sigma_i^2)\delta_{ij}$.

By differentiating $\chi^2$ with respect to each of the $a_r$ and setting these differentials to zero, we obtain $n$ equations—the *normal equations*—which can be solved to find the $\hat{\mathbf{a}}$, the least squares estimates for $\mathbf{a}$. If $f(x; \mathbf{a})$ is *linear* in the $a_r$, then the equations are linear, and can be solved exactly.

Be careful about this. It is linearity in the $a_r$ that matters, not linearity in $x$. This means a deceptively simple form can be non-linear, and vice versa. For example,

$$f(x;\mathbf{a}) = a_1 + a_2 x + a_3 x^4 + a_4 \sin\sqrt{x}$$

is linear, but not

$$f(x;\mathbf{a}) = a_1 + \sin(x + a_2)$$

A linear function can be written

$$f(x;\mathbf{a}) = \sum_r c_r(x) a_r$$

and then, for the diagonal

$$\chi^2 = \sum_{i=1}^{N} \left[ \frac{y_i - \sum_{r=1}^{n} a_r c_r(x_i)}{\sigma_i} \right]^2.$$

Differentiating this with respect to $a_r$ and setting the result to zero gives the normal equations

$$\sum_i c_r(x_i) \left[ \frac{y_i - \sum_s c_s(x_i)\hat{a}_s}{\sigma_i^2} \right] = 0 \quad \text{for all } r.$$

Introducing another matrix

$$C_{ir} = c_r(x_i)$$

we can rewrite these equations in matrix form

$$\mathbf{f} = \mathbf{Ca}$$

$$\chi^2 = (\tilde{\mathbf{y}} - \tilde{\mathbf{a}}\tilde{\mathbf{C}})\mathbf{V}^{-1}(\mathbf{y} - \mathbf{Ca}) \tag{6.21}$$

and the above set of $n$ normal equations can be written

$$\tilde{\mathbf{C}}\mathbf{V}^{-1}\mathbf{C}\hat{\mathbf{a}} = \tilde{\mathbf{C}}\mathbf{V}^{-1}\mathbf{y}. \tag{6.22}$$

Actually we have only shown this for independent measurements, for which $\mathbf{V}$ is diagonal; it is also true for the more general case. This (a) looks plausible and (b) can be proved by wading through the algebra, if desired.

Note the shape of these matrices. If there are $N$ data points and $n$ coefficients to be found (so $n \leqslant N$ or there is no unique solution), then $\mathbf{y}$ and $\mathbf{a}$ are vectors (column matrices) but of different dimension, i.e. $N$ and $n$ respectively. $\mathbf{V}$ is an $N \times N$ square matrix, and $\mathbf{C}$ is rectangular, $N$ rows and $n$ columns.

You cannot cancel a factor of $\tilde{\mathbf{C}}$ from this matrix equation, as $\mathbf{C}$ is a rectangular matrix, not square, and does not have an inverse. The least squares estimate of **a** is thus

$$\hat{\mathbf{a}} = (\tilde{\mathbf{C}}\mathbf{V}^{-1}\mathbf{C})^{-1}\tilde{\mathbf{C}}\mathbf{V}^{-1}\mathbf{y} \tag{6.23}$$

and this is the desired solution. How accurate is it?

Well, equation 6.23 shows that $\hat{\mathbf{a}}$ is obtained from **y** by multiplying by a matrix $(\tilde{\mathbf{C}}\mathbf{V}^{-1}\mathbf{C})^{-1}\tilde{\mathbf{C}}\mathbf{V}^{-1}$. The matrix form of the law of combination of errors (equation 4.19) says that under such transformations the variance also transforms using the same matrix, i.e.

$$\textit{if } \mathbf{x}' = \mathbf{M}\mathbf{x} \textit{ then } \mathbf{V}(x') = \mathbf{M}\mathbf{V}(x)\tilde{\mathbf{M}}.$$

So we can immediately write the variance of our estimator $\hat{\mathbf{a}}$ as

$$\mathbf{V}(\hat{\mathbf{a}}) = \mathbf{M}\mathbf{V}(\mathbf{y})\tilde{\mathbf{M}}$$

where

$$\mathbf{M} = [\tilde{\mathbf{C}}\mathbf{V}(\mathbf{y})^{-1}\mathbf{C}]^{-1}\tilde{\mathbf{C}}\mathbf{V}(\mathbf{y})^{-1}.$$

Note that we now have to distinguish between the variance matrix for the measurements **V(y)** and the variance matrix $\mathbf{V}(\hat{\mathbf{a}})$ for the results. By a special dispensation of providence, this whole thing can be simplified (remembering that $\mathbf{V} = \tilde{\mathbf{V}}$ and $\widetilde{\mathbf{AB}} = \tilde{\mathbf{B}}\tilde{\mathbf{A}}$) to

$$\mathbf{V}(\hat{\mathbf{a}}) = [\tilde{\mathbf{C}}\mathbf{V}(\mathbf{y})^{-1}\mathbf{C}]^{-1}. \tag{6.24}$$

This is what a statistician would call the covariance matrix for the estimator $\hat{\mathbf{a}}$, but is more familiar as the 'error on the results of the least squares fit'. Note that it comes free, as you have to evaluate this matrix to solve equation 6.23 anyway (although there may be ways round this, so this statement is not strictly true). These two equations—6.23 and 6.24—contain the complete method of linear least squares.

### ★6.6.1 Straight Line Fit Using Matrices

Use of matrices in least squares can be illustrated by the problem of fitting a straight line, with all measurements equally accurate, already treated in section 6.2. In matrix language, $f(x) = a_0 + a_1 x$ and $\mathbf{V} = \sigma^2\mathbf{I}$, and

$$\mathbf{C} = \begin{pmatrix} 1 & x_1 \\ 1 & x_2 \\ 1 & x_3 \\ \vdots & \vdots \end{pmatrix} \qquad \tilde{\mathbf{C}} = \begin{pmatrix} 1 & 1 & 1 & \cdots \\ x_1 & x_2 & x_3 & \cdots \end{pmatrix}.$$

Equation 6.23 reduces (because of the simple form of **V**) to

$$\hat{\mathbf{a}} = \sigma^2(\tilde{\mathbf{C}}\mathbf{C})^{-1}\frac{1}{\sigma^2}\tilde{\mathbf{C}}\mathbf{y} \tag{6.25}$$

which, putting in the form for **C** and doing some matrix multiplications, gives

$$\hat{\mathbf{a}} = \begin{pmatrix} \hat{a}_0 \\ \hat{a}_1 \end{pmatrix} = \begin{pmatrix} \sum_i 1 & \sum_i x_i \\ \sum_i x_i & \sum_i x_i^2 \end{pmatrix}^{-1} \begin{pmatrix} \sum_i y_i \\ \sum_i x_i y_i \end{pmatrix}.$$

Inversion of this $2 \times 2$ matrix is easy. It can be written

$$\frac{1}{N(\overline{x^2} - \bar{x}^2)} \begin{pmatrix} \overline{x^2} & -\bar{x} \\ -\bar{x} & 1 \end{pmatrix}$$

which gives equations 6.6 and 6.8 directly.

For the errors we need the inverse of the matrix (see equation 6.24):

$$\tilde{\mathbf{C}}\mathbf{V}^{-1}\mathbf{C} = \frac{1}{\sigma^2}\tilde{\mathbf{C}}\mathbf{C}.$$

This has just been evaluated above, so the answer is

$$\mathbf{V}(\hat{\mathbf{a}}) \equiv \begin{pmatrix} V(a_0) & \operatorname{cov}(a_0, a_1) \\ \operatorname{cov}(a_0, a_1) & V(a_1) \end{pmatrix} = \frac{\sigma^2}{N(\overline{x^2} - \bar{x}^2)} \begin{pmatrix} \overline{x^2} & -\bar{x} \\ -\bar{x} & 1 \end{pmatrix}$$

which contains equations 6.10, 6.11, and 6.12.

### ★ 6.6.2 Higher Polynomials

For fitting a parabola, $f(x) = a_0 + a_1 x + a_2 x^2$, **C** is

$$\mathbf{C} = \begin{pmatrix} 1 & x_1 & x_1^2 \\ 1 & x_2 & x_2^2 \\ 1 & x_3 & x_3^2 \\ \vdots & \vdots & \vdots \end{pmatrix}.$$

Equation 6.25 is still true, and gives

$$\hat{\mathbf{a}} = \begin{pmatrix} \hat{a}_0 \\ \hat{a}_1 \\ \hat{a}_2 \end{pmatrix} = \begin{pmatrix} \sum_i 1 & \sum_i x_i & \sum_i x_i^2 \\ \sum_i x_i & \sum_i x_i^2 & \sum_i x_i^3 \\ \sum_i x_i^2 & \sum_i x_i^3 & \sum_i x_i^4 \end{pmatrix}^{-1} \begin{pmatrix} \sum_i y_i \\ \sum_i x_i y_i \\ \sum_i x_i^2 y_i \end{pmatrix}. \tag{6.26}$$

The extension to cubics, quartics, or polynomials of any order is obvious.

Actually it is rather too obvious. This sort of matrix (called a Hankel matrix) has no shortcuts to its inverse, despite the high degree of symmetry and regularity in its structure. Furthermore, when you do invert it numerically, it turns out to have problems with accuracy and rounding.

A cleaner approach is to construct a set of polynomials up to the desired order, which are orthogonal over the measured data points, i.e. for which $\sum_i P_r(x_i)P_s(x_i) = 0$ unless $r = s$. Such a set can always be found for a given set of $x$ measurements (using a method known as the Schmidt orthogonalisation procedure). Then the matrix $\tilde{\mathbf{C}}\mathbf{C}$ is diagonal and the inverse is trivial.

Despite all that, for the benefit of those of you who insist on fitting simple parabolas, out of the goodness of my heart I will give the explicit form for the inverse of the above matrix. It is

$$\frac{1}{N\Delta}\begin{pmatrix} \overline{x^2}\,\overline{x^4}-\overline{x^3}^2 & \overline{x^2}\,\overline{x^3}-\overline{x}\,\overline{x^4} & \overline{x}\,\overline{x^3}-\overline{x^2}^2 \\ \overline{x^2}\,\overline{x^3}-\overline{x}\,\overline{x^4} & \overline{x^4}-\overline{x^2}^2 & \overline{x}\,\overline{x^2}-\overline{x^3} \\ \overline{x}\,\overline{x^3}-\overline{x^2}^2 & \overline{x}\,\overline{x^2}-\overline{x^3} & \overline{x^2}-\overline{x}^2 \end{pmatrix} \tag{6.27}$$

where

$$\Delta = \overline{x^2}\,\overline{x^4} - \overline{x^3}^2\,\overline{x^4} + 2\overline{x}\,\overline{x^2}\,\overline{x^3} - \overline{x^2}^3$$

## ★6.7 NON-LINEAR LEAST SQUARES

If the function $f(x;\mathbf{a})$ is not linear in the $a_r$, you generally have to use an iterative technique to solve the equations. Given a first guess, $\mathbf{a}^\circ$, then the gradients are (for independent measurements)

$$\left.\frac{\partial\chi^2}{\partial a_r}\right|_{\mathbf{a}=\mathbf{a}^\circ} = g_r(\mathbf{a}^\circ) = \sum_i -\frac{2}{\sigma_i^2}[y_i - f(x_i;\mathbf{a}^\circ)]\frac{\partial f(x_i;\mathbf{a}^\circ)}{\partial a_r}. \tag{6.28}$$

You want to find an increment $\delta\mathbf{a}$ such that

$$g_r(\mathbf{a}^\circ + \delta\mathbf{a}) = \left.\frac{\partial\chi^2}{\partial a_r}\right|_{\mathbf{a}=\mathbf{a}^\circ+\delta\mathbf{a}} = 0 \quad \text{for all } r.$$

Do this by expanding (6.28) in a Taylor series, keeping only the zeroth and first order terms

$$g_r(\mathbf{a}^\circ + \delta\mathbf{a}) \approx g_r(\mathbf{a}^\circ) + \sum_s \frac{\partial g_r}{\partial a_s}\delta a_s = g_r(\mathbf{a}^\circ) + \sum_s \frac{\partial^2\chi^2}{\partial a_r \partial a_s}\delta a_s.$$

Writing $f(x_i;\mathbf{a}^\circ)$ as $f_i$ for brevity, and

$$G_{rs} = \frac{\partial^2\chi^2}{\partial a_s \partial a_r} = \sum_i \left(\frac{-2}{\sigma_i^2}\right)\left[-\left(\frac{\partial f_i}{\partial a_r}\right)\left(\frac{\partial f_i}{\partial a_s}\right) + (y_i - f_i)\left(\frac{\partial^2 f_i}{\partial a_r \partial a_s}\right)\right].$$

$\delta\mathbf{a}$ is found from the matrix equation

$$\delta\mathbf{a} = -\mathbf{G}^{-1}\mathbf{g}$$

and then the solution is found by iteration.

In such iterative solutions, having a good first guess is 90% of the battle. The matrices are then constructed and inverted. When the $g_r$ become 'small enough' the iteration ceases and the resulting $\mathbf{a}$ can be taken as the solution. There are many practical details involved, like whether all terms need be reevaluated at every iteration, and what to do if $\chi^2$ gets worse. As usual, the best advice is to find a reliable software library and use one of its packages. Do not be tempted to write your own.

## 6.8 PROBLEMS

*6.1*
A trolley moves along a track with a constant speed, which you need to measure. It passes through the point $d = 0$ at exactly $t = 0$. At certain fixed times, determined by a stroboscope, you photograph the trolley and can thus measure its position with an error of 2mm. The results are:

| Time $t$ (seconds) | 1.0 | 2.0 | 3.0 | 4.0 | 5.0 | 6.0 |
|---|---|---|---|---|---|---|
| Distance $d$ (mm) | 11 | 19 | 33 | 40 | 49 | 61 |

Find the velocity, the error on this determination, and $\chi^2$.

*6.2*
A trolley moves along a track with a constant speed, which you need to measure. It passes through the point $d = 0$ at exactly $t = 0$. At certain fixed distances, determined by sensing devices on the track, the time is measured with an error of 0.1 sec. The results are:

| Time $t$ (seconds) | 1.1 | 2.2 | 2.9 | 4.1 | 5.0 | 5.8 |
|---|---|---|---|---|---|---|
| Distance $d$ (mm) | 10 | 20 | 30 | 40 | 50 | 60 |

Find the velocity, the error on this determination, and $\chi^2$.

*6.3*
You are determining the acceleration due to gravity by switching off an electromagnet to release a ball-bearing and measuring the time $t$ it takes to fall a fixed distance $d$, so $d = \frac{1}{2}gt^2$. The distances are measured precisely, the time with an accuracy of 0.01 sec.

| Time (seconds) | 0.16 | 0.40 | 0.58 | 0.72 | 0.97 |
|---|---|---|---|---|---|
| Distance (metres) | 0.20 | 1.00 | 2.00 | 3.00 | 5.00 |

Calculate the acceleration due to gravity, with the appropriate error, assuming
(a) that the times are as given and
(b) that the field in the magnet takes an unknown but constant time to die away and release the ball-bearing.

Comment on the difference, and on the $\chi^2$ of the two fits.

*6.4*
If it is believed that $y$ is given by $y = ax + b \sin x$, find the least squares estimate for $a$ and $b$. Apply this to the data:

| $x$ (radians) | 0.2 | 0.3 | 0.4 | 0.5 | 0.6 | 0.7 |
|---|---|---|---|---|---|---|
| $y$ | 0.599 | 0.896 | 1.189 | 1.479 | 1.756 | 2.044 |

*6.5*

A decaying radioactive source is observed with a Geiger counter. Readings are taken for a short period (1 minute) at hourly intervals. The number of counts measured is as follows—use them to find the half-life.

| Time (hours) | 0 | 1 | 2 | 3 | 4 | 5 | 6 | 7 | 8 |
|---|---|---|---|---|---|---|---|---|---|
| Counts | 997 | 520 | 265 | 127 | 70 | 35 | 16 | 7 | 3 |

*You would expect, by all the laws of probability, to find a mad grandmother at Cold Comfort Farm, and for once the laws of probability had not done you down and a mad grandmother there was.*

*—Stella Gibbons*

CHAPTER 

# Probability and Confidence

Most statistics courses gloss over the definition of what is meant by *probability*, with at best a short mumble to the effect that there is no universal agreement. The implication is that such details are irrelevancies of concern only to long-haired philosophers, and need not trouble us hard-headed scientists.

This is short-sighted; uncertainty about what we really mean when we calculate probabilities leads to confusion and bodging, particularly on the subject of *confidence levels*. This chapter therefore takes a look at what we mean by the term 'probability' before discussing the serious business of confidence levels. It is important to make this clear, as sloppy thinking and confused arguments in this area arise mainly from changing one's definition of 'probability' in midstream, or, indeed, of not defining it clearly at all.

*'When I use a word', Humpty Dumpty said in a rather scornful tone, 'it means just what I choose it to mean—neither more nor less.'*

*—Lewis Carroll*

## ★7.1 WHAT IS PROBABILITY?

There are four definitions of 'probability' in general use. Each has its own strong points and shortcomings. Sometimes they are complementary, sometimes contradictory. Here they are.

### ★7.1.1 Mathematical Probability

Let $S = \{E_1, E_2, E_3, \ldots\}$ be the set of possible results of an experiment ('Events'). Events are said to be 'mutually exclusive' if it is impossible for them both to occur in a result. For every event $E$ there is a probability $P(E)$ which is a real number satisfying the **Axioms of probability**.

**I** $P(E) \geqslant 0$.
**II** $P(E_1 \text{ or } E_2) = P(E_1) + P(E_2)$ *if* $E_1$ *and* $E_2$ *are mutually exclusive.*
**III** $\sum P(E_i) = 1$, *where the sum is over all possible mutually exclusive events.*

This approach was formulated by A.N. Kolmogorov (see bibliography)—the axioms given here are a simplifed version of his treatment. Today they can be found, in one form or another, in many mathematical textbooks.

From these axioms all the other results of probability theory can be obtained. For example, axiom III leads directly to the fact that $P(\textit{not } E) = 1 - P(E)$, and thus that $P(E) \leqslant 1$. Extending the axioms to results of measuring a continuous variable, where the sample space is infinite, in itself poses no problems.

As far as this goes, this is fine. However, these axioms are entirely free of meaning. They do not tell you anything about what probability *is about.* You can use them to calculate the probability of some complicated sort of event, but what you get at the end is a mere number—you do not know what it means. (Although Kolmogorov uses the frequency theory, discussed in the next section, he does so only as an illustration.) This is the weakness of the definition, but also its strength, as it thereby avoids controversy. The other three definitions that follow do not contradict this one; they merely seek to give it meaning.

### ★7.1.2 Empirical—The Limit of a Frequency

This is the 'orthodox' definition, for scientists at any rate, and the one you are probably used to. An experiment is performed $N$ times, and a certain outcome A occurs in $M$ of these cases. As $N \to \infty$, the ratio $M/N$ tends to a limit, which is defined as the probability $P(A)$ of A. The $N$ performances may be done by repeating the same experiment $N$ times, one after the other, or by simultaneous measurements on $N$ identical experiments. The set of all $N$ cases is called the *collective* or *ensemble*. The frequency definition of probability is largely due to Richard von Mises (see bibliography).

While this idea of probability has a lot going for it, it has features (some would say shortcomings) which should be brought out and not brushed under the carpet.

The 'probability' of an event is not a property just of the particular experiment, but a joint property of the experiment and the collective. This is nicely illustrated by an example (taken from von Mises). It has been found by the German insurance companies that the fraction of their male clients dying when aged 40 is 1.1%. However, we cannot then say that a particular Herr Schmidt has a probability of 1.1% of dying (or, more cheerfully, 98.9% of surviving) between his 40th and 41st birthdays. If data had been collected not from German insured men but from any other sample to which Herr Schmidt belongs (all German men, all men, all Germans, all German insured nonsmoking men, all German hang-glider pilots,...) we would have got a different fraction. All these different numbers can be considered as the 'probability' of his passing away prematurely, so the quoted 'probability' depends not just on the individual but on the collective to which it is considered to belong.

Secondly, the experiment must be repeatable, under identical conditions, with different possible outcomes. This is a very strong requirement. Consider the harmless-sounding phrase. 'It will probably rain tomorrow.' What can we mean by this? There is only one tomorrow, we can wait and see what happens, but we can only do that once. Similar considerations apply to football matches, Russian Roulette, and the Big Bang. Von Mises faces up to this squarely: in his view, such a use of the word 'probability' is unscientific, in the same way that 'work' and 'energy' are often so (mis)used. However, one would think that if the barometer falls, the clouds gather, and satellite pictures show a cold front approaching, then to say that it will probably rain is a meaningful and sensible statement, and that any definition of probability which forbids it is unduly restrictive.

### ★7.1.3 Objective—Propensity

Originally probability was thought of as a hard, objective, quantity with an existence in its own right. C.S. Peirce, in 1910, writes

> I am, then, to define the meaning of the statement that the *probability*, that if a die be thrown from a dice box it will turn up a number divisible by three, is one-third. The statement means that the die has a certain 'would-be'; and to

> say that the die has a certain 'would-be' is to say that it has a property, quite analogous to any *habit* that a man have.

No mention here of collectives; the 'would-be' is an intrinsic property of the die (or, if you insist, the die, the dice-box, and the rest of the apparatus). This view was eclipsed by the frequency interpretation, but in recent years it has been resurrected, particularly by Sir Karl Popper (see bibliography). His argument, briefly, is that, according to quantum mechanics, science only predicts probabilities, not certainties. If probabilities have no objective reality, as their values depend on the collective, chosen at the whim of the calculator, then the particles they describe have no 'real' properties, behaviour, or existence. Disliking this, Popper proposes an objective probability, or *propensity*, which exists in its own right—though its only observable effect is to drive the ordinary frequency limit probability.†

Objective probability seems very reasonable when considering *equally likely cases.* If a symmetric die has six identical faces each must have the same probability of occurrence, i.e. 1 in 6. A perfect coin must have a 50% probability of producing a head or a tail. It seems unnecessary to drag in arguments about infinite collectives.

However, for a continuous variable this breaks down. If an angle $\theta$ is 'random', then symmetry in $\theta$ argues that the probability of its lying between 80 and 90 degrees must be the same as for 10 and 20 degrees. The former covers a range of a $\cos\theta$ roughly four times greater than the latter, so symmetry in $\cos\theta$ argues that its probability must be four times larger. Other functions would give yet more different and incompatible values. Transforming between variables in a non-trivial way will make any symmetric, uniform distribution non-uniform, and there is no natural choice for the best variable. It is perfectly possible to specify that the distributions are uniform in one or the other variable, but this amounts to a definition of the collective, and the objectivity is lost.

### ★ 7.1.4 Subjective Probability

This view of probability is also known as *Bayesian statistics*, and to discuss this we need Bayes' theorem, and for that we need to define *conditional probability.*

The conditional probability $p(a|b)$ is the probability of $a$ *given* that $b$ is true. For example, if a day is chosen at random then $p(\text{Monday}) = 1/7$. However, if we know that it is a weekday, this becomes $p(\text{Monday}|\text{weekday}) = 1/5$. Likewise selecting cards from a peak, $p(\spadesuit 2)$ is

†'We have to visualize the conditions as endowed with a tendency or disposition, or propensity, to produce sequences whose frequencies are equal to the probabilities; which is precisely what the propensity interpretation asserts'. Popper: *The Logic of Scientific Discovery*, 1959, p. 35.

1/52, but $p(\spadesuit 2|\spadesuit)$ is 1/13, and $p(\spadesuit 2|\text{not face card})$ is 1/40. This is just an extension of the mathematical ideas of section 7.1.1, and does not involve any difficulties of interpretation.

Bayes' theorem (from the Rev. Thomas Bayes, in 1763) is in itself uncontroversial and non-heretical. It uses the simple construction:

$$p(a|b)p(b) = p(a \text{ and } b) = p(b|a)p(a)$$

to give

$$p(a|b) = \frac{p(b|a)p(a)}{p(b)}. \tag{7.1}$$

It is often helpful to write $p(b)$, the probability that $b$ will happen whether $a$ is true or not, in the form (where $\bar{a}$ denotes 'not $a$')

$$p(b) = p(b|a)p(a) + p(b|\bar{a})[1 - p(a)] \tag{7.2}$$

*Example A Čerenkov counter*

A beam of mesons, composed of 90% pions and 10% kaons, hits a Cerenkov counter. In principle the counter gives a signal for pions but not for kaons, thereby identifying any particular meson. In practice it is 95% efficient at a giving a signal for pions, and also has a 6% probability of giving an accidental signal from a kaon. If a meson gives a signal, we can use Bayes' theorem to say that the probability of its being a pion is

$$p(\pi|\text{signal}) = \frac{p(\text{signal}|\pi)}{p(\text{signal}|\pi)p(\pi) + p(\text{signal}|K)p(K)} p(\pi)$$

$$= \frac{0.95}{0.95 \times 0.90 + 0.06 \times 0.10} \times 0.90 = 99.3\%$$

The probability of its being a kaon is the complement of this:

$$p(K|\text{signal}) = 0.7\%.$$

If there is no signal, we have

$$p(K|\text{no signal}) = \frac{p(\text{no signal}|K)}{p(\text{no signal}|\pi)p(\pi) + p(\text{no signal}|K)p(K)} p(K)$$

$$= \frac{0.94}{0.05 \times 0.90 + 0.94 \times 0.10} \times 0.10 = 67.6\%.$$

So the presence of a signal indicates an almost certain $\pi$; its absence indicates a probable, but not certain, $K$.

*Although I felt sure that parity would not be violated, there was a possibility that it would be, and it was important to find out. 'Would you bet a hundred dollars against a dollar that parity is not violated?' he asked. 'No. But fifty dollars I will'.*

*—R.P. Feynman*

*quoted in* The Ambidextrous Universe, *by Martin Gardner*

### ★7.1.5 Bayesian Statistics

*Subjective probability*, also known as *Bayesian statistics*, pushes Bayes' theorem further by applying it to statements of the type described as 'unscientific' in the frequency definition. The probability of a theory (e.g. that it will rain tomorrow or that parity is not violated) is considered to be a subjective 'degree of belief'—it can perhaps be measured by seeing what odds the person concerned will offer as a bet. Subsequent experimental evidence then modifies the initial degree of belief, making it stronger or weaker according to whether the results agree or disagree with the predictions of the theory in question. Mathematically, equation 7.1 becomes

$$p(\text{theory}\,|\,\text{result}) = \frac{p(\text{result}\,|\,\text{theory})}{p(\text{result})}\, p(\text{theory}). \tag{7.3}$$

This behaves in a sensible way. If an observed result is forbidden by some theory, i.e. $p(\text{result}\,|\,\text{theory}) = 0$, then the observation of that result 'disproves' the theory i.e. $p(\text{theory}\,|\,\text{result})$ is zero. If the result is said by the theory to be highly unlikely, then observing it will make the degree of belief in the theory smaller. If a result is predicted to have a high probability, then when it happens it strengthens one's belief in the theory, though this is tempered by the $p(\text{result})$ term in the denominator; a result which is likely to happen for other reasons does not provide strong support for a theory that predicts it.

*Example Subjective probability*

In a moment of boredom you take a coin from your pocket and toss it three times. It comes down heads each time. Later that evening you are drinking in a sleazy bar with Honest Harry, the used car salesman, and toss a coin to see which of you pays for the drinks: heads you pay, tails he pays. He wins three calls out of three.

In the first case, three heads is hardly likely to make you suspect the coin is biased (i.e. double-headed). In the second you would get pretty suspicious—at least to the extent of providing your own coin on the fourth round. Yet the mathematics of $(\frac{1}{2})^3 = \frac{1}{8}$ is the same in both cases.

A Bayesian approach can sort out this difference. If you choose a coin at random you might guess that the 'probability' of its being a double-headed phoney is very small—say, one in a million. After three consecutive heads this becomes

$$p(\text{phoney}\,|\,3\text{ heads}) = \frac{1}{0.125 \times 0.999999 + 1 \times 0.000001} \times 10^{-6} = 8 \times 10^{-6}$$

which is still tiny.

On the other hand, the *a priori* probability that 'Honest Harry' is cheating is, though small, not that small. 5% seems a reasonable guess. This gives

$$p(\text{phoney}\,|\,3\text{ heads}) = \frac{1}{0.125 \times 0.95 + 1 \times 0.05} \times 0.05 = 30\%$$

which is a considerable chance.

This is all very well *if* the initial degree of belief is accepted. Herein lies the problem. Any attempt to base the initial belief on guesswork or instinct must be unscientific and unreliable. The only strict way to justify an initial degree of belief is by the equally likely method introduced in section 7.1.3. As we saw, this does not work in a continuous case.

### ★7.1.6 Conclusions on Probability

Thus probability can be considered as the limit of a frequency, as an objective number or as a subjective degree of belief. This has been a very quick look at a very deep subject, and you should be aware that there are serious differences even within these camps. Beware, too, of names: some people refer to the frequency definition of probability as 'objective', the Bayesians call the frequentists 'classical', and the frequentists call the equally likely school 'classical'.

Why have we opened this can of worms? There is no point in arguing the claims of rival schools: you can adopt whatever definition you please, and use arguments about the merits of different definitions as an amusing conversation topic. What matters is that you should be aware of what you are doing, and do not mix up thoughts, ideas, and formulae from the different definitions.

Most scientists, if challenged, would claim to belong to the frequency school. Propensities and Bayesian statistics are strictly unorthodox and heretical. However, although we claim to adopt the frequency definition, in our innermost hearts we probably think of probabilities as objective numbers, and often talk in language appropriate to Bayesian probabilities. In particular, any attempt to interpret the results of an experiment falls into the trap of repeatability.

Suppose you measure the mass of the electron as $520 \pm 10\,\mathrm{keV/c^2}$. This is a clear statement; you have obtained a result of $520\,\mathrm{keV/c^2}$ with an apparatus of known resolution $10\,\mathrm{keV/c^2}$. You may then say, on the basis of your value, that 'the mass of the electron probably lies close to $520\,\mathrm{keV/c^2}$' or even make the more numerically detailed statement that 'the value lies between 510 and 530, with a 68% probability'. Either statement is, in von Mises' view, 'unscientific' and incompatible with your claimed adherence to the frequency definition. The electron has just one mass (it happens to be $511\,\mathrm{keV/c^2}$) and it either lies within your error bar or outside it.

Such statements are really using subjective, Bayesian, arguments. Before the experiment you know nothing about $m_e$, so you consider all possibilities equally likely. Now, having made a measurement $m$ of resolution $\sigma$, so that

$$p(m|m_e) \propto e^{-(m-m_e)^2/2\sigma^2}$$

Bayes' theorem turns this round to say

$$p(m_e|m) \propto e^{-(m-m_e)^2/2\sigma^2}.$$

If you want to do this, then that is fine, but do it with your eyes open. The conclusion rests on the initial uniform distribution which, as stressed earlier, is not automatic. You could have interpreted this as a measure of $m_e^2$, about which you are equally ignorant initially, and assumed that all values of $m_e^2$ (instead of $m_e$) are equally likely. Then you would get a different result.

To discuss such experimental results and confidence levels in the frequency interpretation, one is forced into a slightly contorted view-point. This is described in the next section.

*'That's a great deal to make one word mean',*
*Alice said in a thoughtful tone.*
*'When I make a word do a lot of work like that',*
*said Humpty Dumpty, 'I always pay it extra.'*

—*Lewis Carroll*

*We* warn *the reader that there is no universal*
*convention for the term 'confidence level.'*

—*The Review of Particle Properties, 1986*

## ★7.2 CONFIDENCE LEVELS

Confidence levels appear as a part of descriptive statistics, as ways of describing the spread of a distribution, especially in the tails. We will look at their definition and properties in this context first; they are very basic and simple. In the next section we go on to the more subtle business of their use in estimation, and the results of measurements.

### ★7.2.1 Confidence Levels in Descriptive Statistics

Suppose cereal packets are produced according to a Gaussian distribution of mean 520 g and standard deviation 10 g. The table of the integrated Gaussian, Table 3.2, then tells us that 68% of packets will weigh more than 510 and less than 530 g. So if we say, when challenged by a consumer group, that the weight of a packet lies in the interval 510 to 530 g, we will be correct 68% of the time. We make the statement with 68% confidence. This

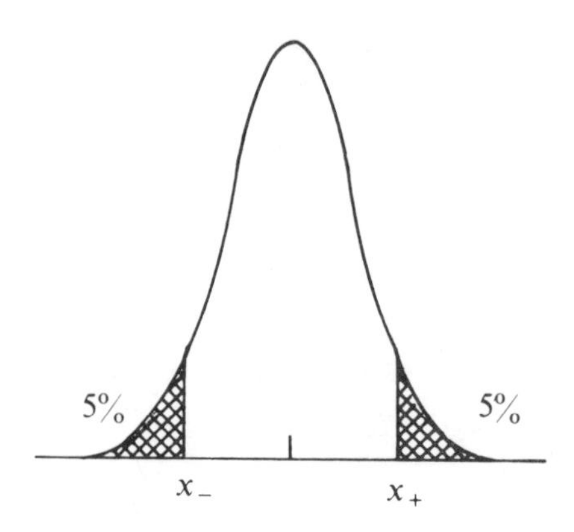


Fig. 7.1. The 90% central confidence interval for a Gaussian distribution.

is a probability according to the standard definition as a frequency limit, as the number of cereal packets coming off the production line is large.

There is a lot of choice about the confidence to quote. Common values are 68% or $1\sigma$, 95.4% ($2\sigma$), 90% ($1.64\sigma$), 95% ($1.96\sigma$) and 99% ($2.58\sigma$). There is a trade-off between a narrow interval and low confidence. You can say with great confidence that the weight lies within very wide limits; if you want to tie it down more precisely the confidence lessens. In practice, 90 and 95% (or $2\sigma$) limits are commonly met with; 99% limits are occasionally used by perfectionists.

For non-Gaussian distributions the correspondence listed above between confidence levels and number of $\sigma$ no longer applies. If someone quotes a '$2\sigma$' result for a non-Gaussian distribution they may mean two standard deviations, or they may in fact mean, rather misleadingly, 95.4%. Care is necessary here.

Having chosen the value, there is still a choice over the range. There are three conventional way of choosing the limits of an interval around the centre. If the limits are $x_-$ and $x_+$, then, for a given confidence level $C$ they obey the requirement

$$\mathrm{Prob}(x_- \leqslant x \leqslant x_+) = \int_{x_-}^{x_+} P(x)\,\mathrm{d}x = C \tag{7.4}$$

and additional requirements as follows:

1. The symmetric interval: $x_-$ and $x_+$ are equidistant from the mean $\mu$, i.e. $x_+ - \mu = \mu - x_-$.
2. The shortest interval: the limits are such that the interval is as short as possible, subject to equation 7.4; i.e. $x_+ - x_-$ is a minimum.

3. The central interval: The probabilities above and below the interval are equal, i.e. $\int_{-\infty}^{x_-} P(x)\,dx = \int_{x_+}^{\infty} P(x)\,dx = (1 - C)/2$.

The central interval is usually the most sensible and the best one to use. However, for the Gaussian distribution (and indeed for any symmetric distribution) the three definitions are all equivalent anyway, so the problem does not often appear.

Two other useful forms are the one-tailed limits, the upper and lower limits. At the stated confidence level, the packet weights lie below the upper limit, i.e.

$$\text{Prob}(x < x_+) = \int_{-\infty}^{x_+} P(x)\,dx = C \tag{7.5}$$

and one does not care what their weights are at low values. Similarly, the weights lie (at the stated confidence level) above the lower limit

$$\text{Prob}(x > x_-) = \int_{x_-}^{\infty} P(x)\,dx = C \tag{7.6}$$

and whether they exceed it by a little or a lot is irrelevant.

*Careful!* It must be emphasized that the upper limit of a 95% central confidence interval, and the 95% upper limit, are not the same thing. The former has 97.5% of the probability content below it and 2.5% above; the latter has 95% below and 5% above.

### ★7.2.2 Confidence Intervals in Estimation

Suppose we want to know the value of a parameter $X$, and have estimated it from the data, giving a result $x$. We know about the resolution of our measurements, and thus $V(x)$ and its square root $\sigma$. The problem is to turn our knowledge of $x$ and $\sigma$ into a statement, of the confidence level type, about the true value $X$.

The naive answer is to turn it round and say '$X$ lies within $x - \sigma$ and $x + \sigma$, with 68% confidence, and within $x - 2\sigma$ and $x + 2\sigma$, with 95% confidence'. However, as described in section 7.1.6, this apparently simple statement is dynamite, containing hidden Bayesian assumptions. Anyone still tempted to think in these terms is invited to consider the following example, which shows that applying probabilities like this is just wrong!

*Example An impossible probability*

The weight of an empty dish is measured as $25.30 \pm 0.14$ g. A sample of powder is placed on the dish, and the combined weight measured as $25.50 \pm 0.14$ g. By subtraction, and combination of errors, the weight of the powder is $0.20 \pm 0.20$ g. This is a perfectly sensible result, though the poor scientist involved should probably find a more accurate balance.

However, look what happens to the probabilities! The naive statement now says that there is a 32% chance of the weight being more than $1\sigma$ from the mean, which is evenly split, making a 16% chance that the weight is negative. This, as Euclid used to say, is absurd.

We will now approach the problem more carefully, using the frequency limit definition of probability. For a particular value of $X$, there is a probability distribution function for $x$: $P(x; X)$. For a conventional measurement of resolution $\sigma$ it is a Gaussian for $x$ with mean $X$ and standard deviation $\sigma$; for a number of (Poisson) events it is the Poisson formula of $x$ given a mean $X$. In general, it presumably peaks at or near $x = X$ and falls off to either side. From it we can construct a confidence interval—let us use a 90% central interval—so that, for a particular value of the real $X$, the value of the measurement $x$ will lie (with 90% probability) within the region $x_-$ to $x_+$. For a different $X$, there are different limits. Thus $x_-$ and $x_+$ can be considered as functions of $X$. This can be nicely shown on a diagram (Figure 7.2). $X$ runs vertically, and a horizontal line at a particular value of $X$ cuts the curves shown, enabling the values of the limits $x_-$ and $x_+$ for this $X$ to be read off on the horizontal axis. The region between the two curves is called the *confidence belt*. The key to these plots is that they are *constructed horizontally* before you ever see the data, using the probability distribution $P(x; X)$, and *read vertically* when you have a measurement.

Now you make an actual measurement $x$. The $x_-$ curve gives the value of $X$ for which $x$ is the appropriate $x_-$. This is the desired *upper* limit $X_+$. This is not saying that $X$ has a 5% probability of exceeding $X_+$—a statement previously condemned as naive and even heretical. It means that if the real $X$ is $X_+$ or greater, then the probability of getting a measurement smaller than

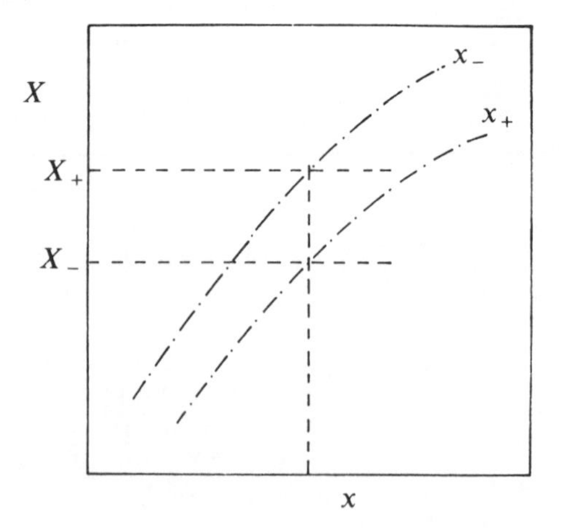


Fig. 7.2. A confidence diagram.

this is 5% or less. Likewise the value of $X$ for which our value of $x$ is $x_+$ is the lower limit $X_-$. We therefore quote the 90% confidence interval for the true value $X$ as the range $X_-$ to $X_+$.

When $X_-$ and $X_+$ are constructed in this way, we can still say the true value of $X$ lies in the range $X_- \leqslant X \leqslant X_+$ with 90% probability. This looks like a statement about $X$, but in fact it is a statement about $X_+$ and $X_-$. Suppose the true value of $X$ is $X_0$, and it is measured many times. The many (different) measurements will, by construction, lie within the range $x_-$ to $x_+$ (inclusive) as evaluated for $X_0$ in 90% of all cases, while the other 10% will not. Points inside the belt are within their horizontal limits ($x_-$ and $x_+$ for this $X$) and also their vertical limits ($X_-$ and $X_+$ for this $x$). Points outside violate both bounds. The 90% of measurements within the $x_-$ to $x_+$ range also have $X_0$ in the range $X_-$ to $X_+$. So $X_0$ will lie within the limits $X_-$ to $X_+$ with a probability of 90%. Although a particular statement obtained at, say, a 90% confidence level (e.g. $m_e$ lies within 510 and 515 keV/C$^2$) is either right or wrong, if you take a large number of such statements then 90% of them will be true.

### ★7.2.3 Confidence Levels from Gaussians

For a Gaussian distribution this conversion from horizontal to vertical is very simple—indeed, deceptively so. Given a measurement $x$ of the mean $X$, and knowing $\sigma$, a 90% confidence interval for $X$ requires the values $X_-$ and $X_+$ such that (looking at the confidence diagram)

$$\int_x^{\infty} \frac{1}{\sigma\sqrt{2\pi}} e^{-(x'-X_-)^2/2\sigma^2}\, dx' = 0.05 = \int_{-\infty}^{x} \frac{1}{\sigma\sqrt{2\pi}} e^{-(x'-X_+)^2/2\sigma^2}\, dx.$$

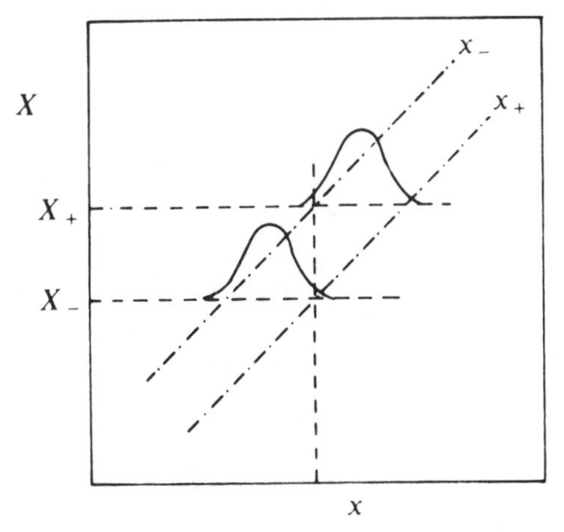


Fig. 7.3. The confidence diagram for a Gaussian.

The equation for $X_-$ requires that $x$ lies some number of standard deviations (in this case 1.64) above $X_-$. This is the same as saying that $X_-$ must lie the same number of $\sigma$ below the measured $x$, which can be written in the form

$$\int_{-\infty}^{X_-} \frac{1}{\sigma\sqrt{2\pi}} e^{-(x'-x)^2/2\sigma^2}\, dx' = 0.05$$

and such confidence limits can be found for Gaussian estimators using the usual table of the Gaussian integral. The curves in Figure 7.2 become, in Figure 7.3, two straight lines with unit gradient, $x_\pm = X \pm n\sigma$ when constructed horizontally, $X_\pm = x \pm n\sigma$ when read vertically, where $n$ is 1 for 68% confidence, 1.64 for 90% confidence, etc., as given by the table of integrated Gaussians. The confidence interval for $X$ obtained from a measurement $x$ is merely $x \pm n\sigma$.

*In fact, so it seems to us, confidence-interval theory has the defect of its principal virtue: it attains its generality at the price of being unable to incorporate prior knowledge into its statements.*

—*Kendall and Stuart*

### ★7.2.4 Measurement of a Constrained Quantity

Now consider the case where we know that there are definite limits on $X$, which it would be physically impossible to exceed. Take the mass of an object as an example: irrespective of any measurement, it has to be positive.

We will use a $2\sigma$ (95.4%) central interval as an illustration. The true mass has some positive value—let suppose it is 0.1 g. This is measured with a resolution of 0.2 g, so a measurement $x$ gives a confidence interval $x \pm 0.4$ g.

There is a 2.3% probability that the measurement will be greater than 0.5 g. From this we will quote limits which are wrong, but the 2.3% probability for this is part of the game and acceptable.

If the measurement lies in the range 0.4 to 0.5, we will quote similar limits, and this time they will be true. If it falls a bit below 0.4, say to 0.3, the limits are $-0.1$ to 0.7; the lower limit can be modified to 0.0 on the basis of common sense, and they are still true. This continues all the way down to a measurement of $-0.3$.

Below $-0.3$ we again make statements which are false, this time because the upper limit will be less than the true value of 0.1. Again the small 2.3% probability is acceptable.

However, if we get a measurement of $-0.5$, we have to quote a range of

$-0.9$ to $-0.1$. This is patently ridiculous. In using a 95.4% confidence level we know that 4.6% of our statements will be untrue, and accept the odds. We now have independent evidence that this particular statement is one of those 4.6%. In such a case, we would be pretty stupid to make it. However, our confidence level approach based on the frequency distribution can tell us nothing more.

That such measurements give nonsensical limits is obvious. More dangerous are measurements like $-0.39$. The $2\sigma$ limits are then $-0.79$ to 0.01. Changing the lower limit from $-0.79$ to 0.0 is permissible, in that it cannot alter the truth or falsehood of the statement. This gives a very narrow interval of 0.00 to 0.01, with 95.4% confidence. To make such a statement is strictly true, and at the same time totally dishonest. Once your quoted confidence interval covers a region of impossible values, you are in trouble.

If you get in a hole like this, Bayesian statistics provides the only means of escape. When faced with a Gaussian measurement $x$, of a true value $X$, the Bayesian does not construct any confidence diagram, but invokes equation 7.3. In this equation the conditional $p(\text{result}|\text{theory})$ is just the Gaussian distribution probability density for a measurement $x$ arising from a true $X$ with resolution $\sigma$. The $p(\text{result})$ in the denominator does not matter as it is taken care of in the final normalisation. $p(\text{theory})$ represents the intrinsic probability distribution for $X$. Normally this is handled by a rather disingenuous assumption of complete ignorance: nothing is known about $X$, so all values are equally likely, so the initial $p(X)$ is uniform and constant. Taking care of the normalisation gives

$$p(X|x) = \frac{e^{-(x-X)^2/2\sigma^2}}{\sigma\sqrt{2\pi}}$$

and confidence levels for $X$ can be constructed as desired, as described in section 7.2.1. They are (because of the symmetry in the Gaussian between $X$ and $x$) exactly the same as those we obtained using the frequency method.

Our extra knowledge—that $X$ must be positive—is easily incorporated. The initial $p(X)$ is now a step function, zero for $X < 0$ and constant for $X > 0$. Equation 7.3 gives, after normalisation,

$$p(X|x) = \frac{e^{-(x-X)^2/2\sigma 2}}{\int_0^\infty e^{-(x-X')^2/2\sigma^2}\,dX'} \qquad (x > 0). \tag{7.7}$$

Confidence levels can be produced from this as desired, using Table 3.3. The distribution is now non-symmetric so there is a choice between the symmetric, shortest, and central interval.

*Example Bayesian approach to confidence*
A mass is measured as $-0.5 \pm 0.2$ g. The integral in the denominator of equation 7.7 is 0.0062, from the probability of exceeding $2.5\sigma$ in Table 3.3. This table also gives the probability of exceeding $3.24\sigma$ as 0.0006, which is 10% of the previous figure. So the 90% confidence upper limit is $-0.5 + 3.24 \times 0.2 = 0.15$ g.

Although this usage is probably the only way to make meaningful statements from such results, you do so in the knowledge that had you used another variable—$X^2$ or $\sqrt{X}$ or $1/X$—the resulting limits would be incompatible. Your assumption of complete ignorance means different things when applied to different forms of the same basic variable.

### ★7.2.5 Binomial Confidence Intervals

For the binomial distribution the observed variable (call it $r$) is discrete, whereas the 'true' value (call it $R$) is continuous. For discrete variables the integrals in equations 7.4, 7.5, and 7.6 are replaced by summations. The subtle difference in the inequality signs now matters: the two-tailed form (equation 7.4) is inclusive ($r$ lies within the range $r_-$ to $r_+$), and the terms for $r_-$ and $r_+$ are included in the sum, but the one-tailed intervals (equations 7.5 and 7.6) are exclusive ($r$ is less than $r_+, \ldots$) and the term for $r_+$ or $r_-$ is excluded.

Wishing to form, say, a 95% central confidence interval for a given $R$ it will not in general be possible to choose an $r_+$ such that $\sum_0^{r_+} P(r; R) = 0.975$. For safety we round $r_+$ up, and select it such that $\sum_0^{r_+} P(r; R) \geqslant 0.975$, and similarly round $r_-$ down. This means that our final statement will be true at least 95% of the time, and possibly more.

The two confidence diagram curves become staircase-like, as the horizontal coordinate is discrete. Confidence limits can be constructed from the belt as before, with summations replacing integrals. Care over the detail of the definitions is required when fixing the limits of the sums. Thus if $m$ successes are found in $n$ binomial trials, limits on the individual probability $p$ are given by finding $p_-$ and $p_+$ such that (using the 95% central limits as an example)

$$\sum_{r=m+1}^{n} P(r; p_+, n) = 0.975 \qquad \sum_{r=0}^{m-1} P(r; p_-, n) = 0.975.$$

These are known as the *Clopper–Pearson confidence limits.*

*Example A binomial confidence interval*
In a sample of 20 fizzgigs, 4 are obloid. What are the 95% confidence limits on the proportion of obloid fizzgigs?

The lower limit is given by $\sum_0^3 P(r; p_-, 20) = 0.975$.

Trial and error show that for $p = 0.057$, the probabilities of 0 to 3 successes are 0.307, 0.373, 0.216, and 0.079, which sum to 0.975.

The upper limit is given by $\sum_5^{20} P(r; p_+, 20) = 0.975$, which is easier to handle as $\sum_0^4 P(r; p_+, 20) = 0.025$.

For $p = 0.437$, the probabilities of 0 to 4 successes are 0.00001, 0.0002, 0.001, and 0.005, and 0.018, which sum to 0.025.

The limits are thus, with 95% confidence, 0.057 to 0.437.

### ★ 7.2.6 Poisson Confidence Intervals

If $n$ events are observed from a Poisson process of unknown mean $N$, the 90% upper limit (for example) is the value $N_+$ such that

$$\sum_{r=n+1}^{\infty} P(r; N_+) = 0.90 \tag{7.8a}$$

or, equivalently,

$$\sum_{r=0}^{n} P(r; N_+) = 0.10. \tag{7.8b}$$

In English, this means: if the true value of $N$ is really $N_+$, the probability of getting a result $n$ which is this small (or smaller) is only 10%, and for $N$ larger than $N_+$ it is even smaller. Thus we say we are '90% confident' that $N$ is not greater than $N_+$, and averaging over many such statements we will be right 9 times out of 10.

Likewise for the 90% lower limit you require $N_-$ such that

$$\sum_{r=0}^{n-1} P(r; N_-) = 0.90. \tag{7.9}$$

These equations for $N_+$ and $N_-$ (which, by the way, are real numbers, not integers) can be solved by iteration. Some are given in the following table.

TABLE 7.1.
SOME POISSON LIMITS

| | Upper | | | Lower | | |
|---|---|---|---|---|---|---|
| | 90% | 95% | 99% | 90% | 95% | 99% |
| $n = 0$ | 2.30 | 3.00 | 4.61 | — | — | — |
| 1 | 3.89 | 4.74 | 6.64 | 0.11 | 0.05 | 0.01 |
| 2 | 5.32 | 6.30 | 8.41 | 0.53 | 0.36 | 0.15 |
| 3 | 6.68 | 7.75 | 10.05 | 1.10 | 0.82 | 0.44 |
| 4 | 7.99 | 9.15 | 11.60 | 1.74 | 1.37 | 0.82 |
| 5 | 9.27 | 10.51 | 13.11 | 2.43 | 1.97 | 1.28 |
| 6 | 10.53 | 11.84 | 14.57 | 3.15 | 2.61 | 1.79 |
| 7 | 11.77 | 13.15 | 16.00 | 3.89 | 3.29 | 2.33 |
| 8 | 12.99 | 14.43 | 17.40 | 4.66 | 3.98 | 2.91 |
| 9 | 14.21 | 15.71 | 18.78 | 5.43 | 4.70 | 3.51 |
| 10 | 15.41 | 16.96 | 20.14 | 6.22 | 5.43 | 4.13 |

Note that if no events are observed, this can give an upper limit on the ideal number, but no lower limit.

### ★7.2.7 Several Variables—Confidence Regions

If two (or more) variables are being estimated simultaneously then putting confidence limits on both of them is, in the words of Kendall and Stuart, 'a matter of very considerable difficulty'. One may have to be satisfied with the establishment of a *confidence region* in the parameter space, within which the true parameters lie (with a certain confidence). This is very relevant to maximum likelihood estimation, which provides a natural framework for the estimation of several variables, as discussed in section 5.3.4.

Consider first an ML estimate of a single parameter. As shown in section 5.3.3, in the large $N$ limit the values of the parameter $a$ at which the log likelihood function is 0.5 less than its peak value are the 'one sigma' limits of the estimate. These can be taken as Gaussian measurements and treated as in section 7.2.3, where the 'one sigma' limits were shown to give the 68% confidence interval. Thus the range of $a$ for which the log likelihood in $L$ is within 0.5 of its peak value constitutes the 68% confidence interval. As before, by the argument of invariance, it can also plausibly be taken as such at small $N$.

For more than one parameter the likelihood function is harder to plot. In the large $N$ limit, surfaces of constant probability are ellipses for two parameters, hyperellipses for more than two. For small $N$ the surfaces are more complicated, but still exist and can, for two parameters, be displayed. One can thus present the ellipse (or whatever) at which $\ln L$ falls off by 0.5, and say that the true parameter values lie within it, at some confidence level.

However, this level is no longer 68%. We have moved from the Gaussian to the multidimensional Gaussian; large values of the exponent are relatively more likely, and are given by the $\chi^2$ distribution, with number of degrees of freedom equal to the number of parameters. So for two parameters the 'one sigma' confidence region gives the 39% confidence region, and for more parameters the level is even less. For a given number of parameters and desired confidence level, the value of $\chi^2$ is found from tables (such as Table 8.1), and the boundary of the confidence region is given by the curve (or surface) at which $\ln L$ falls from its peak value by half this amount. Thus, for example, for two variables the 90% confidence region is given by the parameters for which $\ln L$ is within 2.3 of its maximum value.

## ★7.3 STUDENT'S *t* DISTRIBUTION

When you make a measurement of known resolution—for example, you measure the weight of a ball-bearing to be 13.5 g, using a balance which is known to have a resolution of 0.1 g—then you quote the answer with its resolution (i.e. $13.5 \pm 0.1$ g), and the statement is interpreted as a confidence interval for a Gaussian distribution, as discussed in section 7.2.3.

This is fine provided you know the resolution—when your balance comes with a convenient label on it telling you its accuracy. Of course it often does, and usually when making measurements you have established the performance of the apparatus. But sometimes this is not the case. (This is particularly true in the social sciences, where dispersion arises due to a spread in the basic data sample, rather than from measurement. So Student's $t$ is a topic more familiar to doctors and economists than physicists and chemists.)

What do you do then? You have to take several measurements and look at the spread. A single measurement gives you an honest estimate, but tells you nothing about the accuracy. $\sigma$ is not known *a priori*, but has to be estimated from a sample of several values: we do not have the true value $\sigma$, but only the estimate $\hat{\sigma}$. If $\mu$ is known we use (cf. equation 5.12)

$$\hat{\sigma} = \sqrt{\overline{(x-\mu)^2}}. \tag{7.10}$$

If $\mu$ is unknown we use (cf. equation 5.14)

$$\hat{\sigma} = s = \sqrt{\frac{N}{N-1}\overline{(x-\bar{x})^2}}. \tag{7.11}$$

The second case is more usual, but not universal.

Instead of the variable $(x-\mu)/\sigma$, which is distributed according to a unit Gaussian (i.e. it has mean zero and standard deviation unity), we have to deal with the variable

$$t = \frac{x-\mu}{\hat{\sigma}}. \tag{7.12}$$

$t$ is *not* normally distributed with unit variance, as it would be if $\hat{\sigma}$ were equal to $\sigma$; the significance of a given deviation between an $x$ and $\mu$ is less when $\hat{\sigma}$ is used in place of $\sigma$, because of the additional uncertainty in $\hat{\sigma}$. In practice, especially for small $N$, it is rather a poor estimate of $\sigma$.

$t$ is described by a distribution called *Student's t distribution*, after its discoverer William Gossett, who wrote under the pen name of 'Student'. Writing

$$t = \frac{(x-\mu)/\sigma}{\hat{\sigma}/\sigma}$$

you can see that $t$ is a unit Gaussian divided by a denominator which is (looking at equation. 7.10 or 7.11) the square root of a $\chi^2$ sum. Our ignorance of $\sigma$ in the numerator cancels our ignorance of $\sigma$ in the denominator, enabling $t$ to contain only the observed quantities $x$ and $\hat{\sigma}$. $n$, the number of degrees of freedom in the $\chi^2$, is $N$ if equation 7.10 is used and $N-1$ for equation 7.11.

The distribution function for $t$ is

$$f(t;n) = \frac{\Gamma((n+1)/2)}{\sqrt{n\pi}\,\Gamma(n/2)} \frac{1}{(1+(t^2/n))^{(n+1)/2}}. \tag{7.13}$$

A proof is given in a later section—meanwhile, Figure 7.4 shows some pictures of it.

This curve is like a Gaussian, and indeed for large $n$ it tends to the unit Gaussian, but the tails at the side are larger, especially for small $n$. The difference between Student's $t$ and the normal distribution becomes important when considering small $n$, and/or large deviations from the centre.

Its use in practice is more complicated than the Guassian because of this extra parameter $n$. Full tables like 3.2 and 3.3 are impossibly cumbersome. To find the significance, you consult tables of critical values (Table 7.2).

One important use of Student's $t$ addresses the problem of a sample with some mean $\bar{x}$ and sample variance $s$, asking what one can say about the true mean $\mu$.

This involves an extra factor of $\sqrt{N}$, as we are discussing the standard error on the mean (cf. section 4.2.1) rather than the error $\sigma$ of the measurements. Now, $(\bar{x}-\mu)/(\sigma/\sqrt{N})$ is unit Gaussian, so the appropriate $t$ is

$$\frac{(\bar{x}-\mu)/(\sigma/\sqrt{N})}{\hat{\sigma}/\sigma} = \frac{\bar{x}-\mu}{\hat{\sigma}/\sqrt{N}}.$$

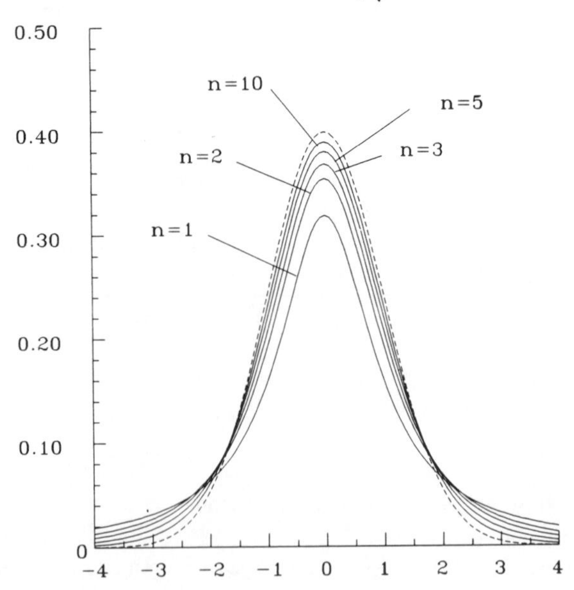


Fig. 7.4. Student's $t$ for $n = 1, 2, 3, 5$, and 10, and the unit Gaussian for comparison.

One forms this, and then consults Table 7.2.

TABLE 7.2
CRITICAL VALUES OF $t$
For various values of confidence levels and $n$

| Confidence (2 tailed) | 60% | 80% | 90% | 95% | 98% | 99% |
|---|---|---|---|---|---|---|
| (1 tailed) | 80% | 90% | 95% | 97.5% | 99% | 99.5% |
| $n=1$ | 1.376 | 3.078 | 6.314 | 12.706 | 31.820 | 63.651 |
| 2 | 1.061 | 1.886 | 2.920 | 4.303 | 6.965 | 9.925 |
| 3 | 0.978 | 1.638 | 2.353 | 3.182 | 4.541 | 5.841 |
| 4 | 0.941 | 1.533 | 2.132 | 2.776 | 3.747 | 4.604 |
| 5 | 0.920 | 1.476 | 2.015 | 2.571 | 3.365 | 4.032 |
| 6 | 0.906 | 1.440 | 1.943 | 2.447 | 3.143 | 3.707 |
| 7 | 0.896 | 1.415 | 1.895 | 2.365 | 2.998 | 3.499 |
| 8 | 0.889 | 1.397 | 1.860 | 2.306 | 2.896 | 3.355 |
| 9 | 0.883 | 1.383 | 1.833 | 2.262 | 2.821 | 3.250 |
| 10 | 0.879 | 1.372 | 1.812 | 2.228 | 2.764 | 3.169 |
| 11 | 0.876 | 1.363 | 1.796 | 2.201 | 2.718 | 3.106 |
| 12 | 0.873 | 1.356 | 1.782 | 2.179 | 2.681 | 3.055 |
| 13 | 0.870 | 1.350 | 1.771 | 2.160 | 2.650 | 3.012 |
| 14 | 0.868 | 1.345 | 1.761 | 2.145 | 2.624 | 2.977 |
| 15 | 0.866 | 1.341 | 1.753 | 2.131 | 2.602 | 2.947 |
| 16 | 0.865 | 1.337 | 1.746 | 2.120 | 2.583 | 2.921 |
| 17 | 0.863 | 1.333 | 1.740 | 2.110 | 2.567 | 2.898 |
| 18 | 0.682 | 1.330 | 1.734 | 2.101 | 2.552 | 2.878 |
| 19 | 0.861 | 1.328 | 1.729 | 2.093 | 2.539 | 2.861 |
| 20 | 0.860 | 1.325 | 1.725 | 2.086 | 2.528 | 2.845 |
| 21 | 0.859 | 1.323 | 1.721 | 2.080 | 2.518 | 2.831 |
| 22 | 0.858 | 1.321 | 1.717 | 2.074 | 2.508 | 2.819 |
| 23 | 0.858 | 1.319 | 1.714 | 2.069 | 2.500 | 2.807 |
| 24 | 0.857 | 1.318 | 1.711 | 2.064 | 2.492 | 2.797 |
| $\infty$ | 0.842 | 1.282 | 1.645 | 1.960 | 2.326 | 2.576 |

*Example Odd one out?*
Four values $\{3.9, 4.5, 5.5, 6.1\}$ are drawn from a sample whose mean is known to be 4.9. Another value of 7.3 is obtained—is it likely to belong to this sample?

$\hat{\sigma}$ can be calculated as 0.86. The deviation of a point at 7.3 is $2.8\hat{\sigma}$ (as opposed to $2.8\sigma$), so $t$ is 2.8. It can be seen from Table 7.2 that for $n=4$ the probability of such a deviation is about 95%. This is probably acceptable; it only has the significance of a two standard deviation discrepancy.

*Example Confidence intervals using Student's t*
A test of 25 professors shows an average IQ of 128, with an $s$ of 15. What are the 95% confidence limits on the true value of the average IQ of all professors?

Assuming these 25 are a fairly chosen random sample, the estimated error on the mean is $15/\sqrt{25} = 3$.

If this were Gaussian, we would use $\pm 1.96\sigma$ and get limits of 122.1 to 133.9. Instead we have to use Student's $t$: the critical $t$ for 24 degrees of freedom is 2.06 (rather than 1.96). So the confidence levels are 121.8 and 134.2—a slightly broader band.

Further uses of Student's $t$ are developed in Chapter 8.

### ★7.3.1 Proof of the Formula for Student's $t$

$t$ is defined as $(x - \mu)/\hat{\sigma}$.

Consider $x$ first. It is normally distributed with mean $\mu$ and standard deviation $\sigma$; therefore $y = (x - \mu)/\sigma$ is normally distributed with mean 0 and standard deviation 1.

Now for $\hat{\sigma}$. The estimate $\hat{\sigma}$ is obtained from

$$\hat{\sigma}^2 = \frac{1}{N-1}\sum_i (x_i - \bar{x})^2$$

or

$$\hat{\sigma}^2 = \frac{1}{N}\sum_i (x_i - \mu)^2$$

so that $n\hat{\sigma}^2/\sigma^2$ is just $\chi^2$, and will be distributed according to the $\chi^2$ distribution with $n$ degrees of freedom, where $n$ is $N$ or $N-1$ or whatever, as appropriate.

So by writing $t$ as

$$t = \frac{x-\mu}{\hat{\sigma}} = \frac{(x-\mu)/\sigma}{\sqrt{(n\hat{\sigma}^2/\sigma^2)/n}} = \frac{y}{\sqrt{\chi^2/n}}$$

we know the distribution in terms of $y$ and $\chi^2$ – it is (equation 6.18)

$$F(y, \chi^2; n) = \frac{e^{-y^2/2}}{\sqrt{2\pi}} \frac{(\chi^2)^{n/2-1} e^{-\chi^2/2}}{2^{n/2}\Gamma(n/2)}.$$

The $y$ can be transformed to a $t$, as $y = t\sqrt{\chi^2/n}$. We pick up a $(\partial y/\partial t) = \sqrt{\chi^2/n}$ term, getting

$$F(t, \chi^2; n) = \frac{1}{\sqrt{2n\pi 2^{n/2}\Gamma(n/2)}} \chi^{n-1} e^{-\chi^2(1+t^2/n)/2}.$$

This can now be integrated over all $\chi^2$, using the standard integral (which can be proved by induction)

$$\int_0^\infty x^m e^{-ax} \, dx = \frac{\Gamma(m+1)}{a^{m+1}}$$

to give the desired result

$$f(t;n) = \frac{\Gamma(n+\frac{1}{2})}{\sqrt{\pi n}\,\Gamma(n/2)} \frac{1}{(1+t^2/n)^{(n+1)/2}}.$$

## ★7.4 PROBLEMS

*7.1*

Laplace argued that when two coins were tossed, there were three possible results, namely two heads, two tails, or one of each, so the probability of each must be 1/3. How would you convince him that he was wrong?

*7.2*

Mongolian swamp fever is such a rare disease that a doctor only expects to meet it once in every 10 000 patients. It always produces spots and acute lethargy in a patient; usually (i.e. 60% of cases) they suffer from a raging thirst, and occasionally (20% of cases) from violent sneezes. These symptoms can arise from other causes: specifically, of patients that do not have this disease. 3% have spots, 10% are lethargic, 2% thirsty, and 5% complain of sneezing. These four probabilities are independent.

Show that if you go to the doctor with all these symptoms, the probability of your having Mongolian swamp fever is 80%, and that if you have them all except sneezing the probability is 46%.

*7.3*

A sample of $n$ trials results in $n_s$ successes and $n_f$ failures. Show that the confidence interval for the ratio $n_s/n_f$ is from $P_-/(1-P_-)$ to $P_+/(1-P_+)$, where $P_\pm$ are the limits, at the appropriate confidence level, as discussed in section 7.2.5.

*7.4*

Show that if eight trials produce four successes and four failures, the 90% confidence interval for the intrinsic probability $p$ is from 19 to 81%. Compare these results with those obtained from a Gaussian approximation to the binomial.

*7.5*

Studying the decay of protons (a rare process!) seven events are observed in 1 year in a sample of 1 000 000 kilograms of hydrogen. Give the 90% confidence interval for the number of decays, and thus for the half-life of the proton, assuming there is no background.

*7.6*

Repeat the above problem, given that the expected background from other random processes is three events per year.

*7.7*

Repeat the above problem, given that the expected background from other random processes is eight events per year.

*7.8*

The usual metal content of a certain ore is 5.6%. Samples of ore from a possible mining site are assayed and have metal contents of 6, 7, 9, and 8%. The average content is better than usual, but is this significant?

*He either fears his fate too much*
*Or his deserts are small*
*Who puts it not unto the touch*
*To win or lose it all.*

—*Montrose*

CHAPTER

# Taking Decisions

Sometimes the information you want from the data is not a number, but the yes-or-no answer to a factual question. You ask not 'What is the straight line fit for $y$ against $x$?' but 'Does $y$ increase with $x$?'; not 'What is the strength of the effect?' but 'Is the effect present?'; not 'What are the values of $a$ and $b$?' but 'Do $a$ and $b$ have the same value?'.

For the answers to be simple, the question posed must be precisely specified. This is done by expressing it as an assertion that some *hypothesis* is true. You then construct a (numerical) *test* and apply it to the data, and the hypothesis is *accepted* or *rejected* depending on the result of the test. Accordingly, this branch of statistics is called *hypothesis testing*.

In this chapter we first go through some general considerations, and then focus on some particular applications: *interpretation of experiments*, *goodness of fit*, the *two-sample problem* in its various forms, and some analyses for *several samples*.

*Hypotheses non fingo*

*—Sir Isaac Newton*

## 8.1 HYPOTHESIS TESTING

When you want to use some data to give the answer to a question, the first step is to formulate the question precisely by expressing it as a hypothesis. Then you consider the consequences of that hypothesis, and choose a suitable test to apply to the data. From the result of the test you accept or reject the hypothesis according to prearranged criteria. This cannot be infallible, and there is always a chance of getting the wrong answer, so you try and reduce the chance of such a mistake to a level which you consider reasonable.

### 8.1.1 Hypotheses

Sometimes a hypothesis is *simple*, in that it specifies the probability distribution completely. For example, "These data are drawn from a Poisson distribution of mean 3.4' or 'The new treatment has identical effects to the old'. On the other hand, it may be *composite*, like 'These data are drawn from a Poisson distribution of mean greater than 4' or 'The new treatment is an improvement on the old'.

Often one has to consider the alternative to the proposed hypothesis. If the data are not Poisson with a mean of 3.4, does this mean that they are Poisson with some other given mean, or some other unspecified mean, or not Poisson at all? The *alternative hypothesis* has to be spelt out. As usual, one has to remember the difference between one-tailed *directional* and two-tailed *non-directional* tests.

### 8.1.2 Type I and Type II Errors

As your decision can be 'Yes' or 'No', and as the hypothesis may really be true or false, there are four possible outcomes. In particular there are two separate and different ways of getting the answer wrong. If you *reject* a *true* hypothesis this is called a *type I* error. If you *accept* a *false* hypothesis this is called a *type II* error.

*Example Type I and type II errors*

In the law courts, the accused proclaims the hypothesis that he is innocent. If the jury reject this and wrongly convict him when he is really innocent, that is a type I error. If they accept his hypothesis and let him off when he is really guilty, that is a type II error.

### 8.1.3 Significance

Type I errors are bound to happen sometimes. How often this occurs is determined by the *significance* of the test, which is pretty well under your control. Suppose the test we apply to the data involves the evaluation of some quantity $x$, and our hypothesis is that the probability distribution has a specified form $P_H(x)$. Then we divide the range of $x$ into the *acceptance region* where $P_H(x)$ is (relatively) large and the *rejection region* where $P_H(x)$ is small, see Figure 8.1. If the actual $x$-falls in the acceptance region then we take the plunge and accept the hypothesis as true (provisionally, at any rate), but if it falls in the rejection region we steel our hearts and reject it.

The probability of rejecting the hypothesis when in fact it is true (type I error) is given by integrating $P_H(x)$ over the rejection region. This is called the *significance* of the test—the reason for the name will appear in due course. It is often denoted by $\alpha$. By choosing the acceptance and rejection regions one has control over this: in a good test $\alpha$ should be small—5 or 1%, say. The smaller the value of $\alpha$, the greater the significance of the test.

More strictly, a test is significant at the $\alpha$ level if the integrated probability of rejecting the true hypothesis is less than or equal to $\alpha$, $\int_R P_H(x)\,dx \leqslant \alpha$. More significant tests embrace the less significant (so a test significant at the 1% level is also significant at the 5% level, as $1 \leqslant 5$). This matters in two important cases. Firstly, if $x$ is discrete, it may not be possible to find a region of $x$ for which the rejection probability is exactly 5%. The extended definition shows that a region for which it is less than 5% will do. Secondly, for a composite hypothesis the integrated probability is unspecified. However, one can find the maximum integrated probability possible under the hypothesis, and if this limit is 5% it cannot (under the hypothesis) be greater, and the extended definition is satisfied.

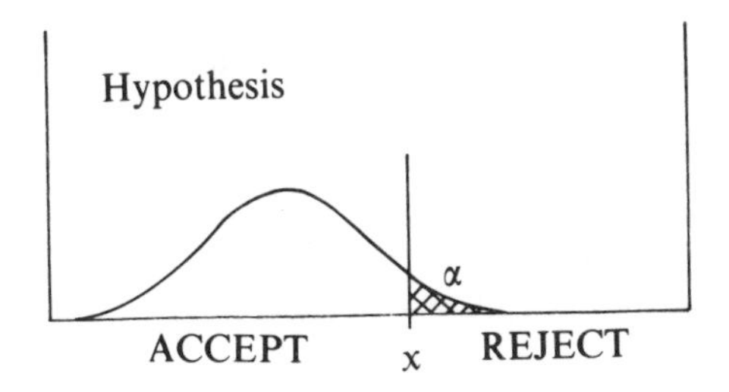


Fig. 8.1. The significance.

### 8.1.4 Power

Now consider type II errors. If the alternative is a simple hypothesis, i.e. with no free parameters, its probability distribution $P_A(x)$ is known. The

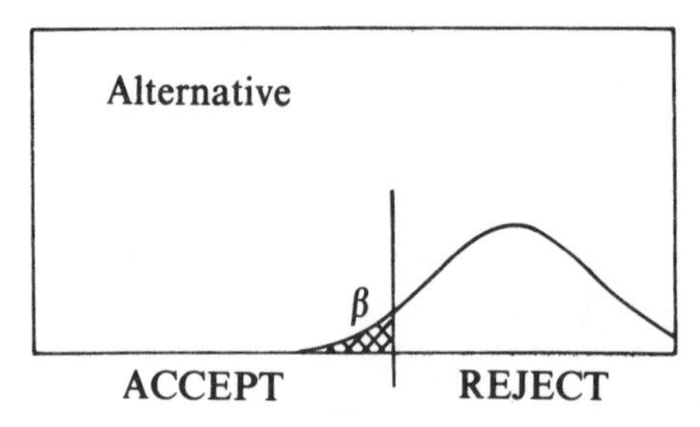


Fig. 8.2. The power.

probability that we will mistakenly accept the hypothesis being tested is $\int P_A(x)\,dx$, where the integral is taken over everywhere outside the rejection region. This is called $\beta$, and $1 - \beta$ is called the *power* of the test, see Figure 8.2. A good test is one for which $\alpha$ and $\beta$ are both small, i.e. the significance is small and the power large. For this to be possible $P_H$ and $P_A$ must be very different.

Generally speaking there is a trade-off between $\alpha$ and $\beta$; making $\alpha$ small will make $\beta$ large. You have to balance the swings and round abouts and decide what you want, bearing in mind the importance of the decision, balancing the penalties of type I and type II errors.

> Some caution is necessary in making deductions even from apparently small $\alpha$. Suppose 20 investigators set out to test a hypothesis which is in fact true. One of them is liable to reject and 'disprove' it at the 5% level, the other 19 do not. This is all right, except that the one with the effect will rush to publish it, whereas the other 19 may not bother.

Do not be overimpressed by the attention and mathematical detail that goes into significance. This is largely due to the fact that it is under control and mathematically tractable. The power is usually far more important, but you can often say very little about it as the alternative hypothesis tends to be vague and unquantifiable. When overawed by a discussion of the significance of some test, it is salutary to remember that you can make a valid but meaningless significance test by choosing a random number $r$ between 0 and 1 and rejecting the hypothesis if $r$ is less than $\alpha$.

### ★8.1.5 The Neyman Pearson Test

The 'best' test is one that makes both $\alpha$ and $\beta$ as small as possible. Such a test can be found if—and only if—the hypothesis and its alternative are simple. It is called a *Neyman Pearson test.*

To make $\beta$ as small as possible for a given $\alpha$, the integral over the chosen rejection region $\int P_A(x)\,dx = 1 - \beta$ must be as large as possible, for a given (small) $\int P_H(x)\,\mathrm{d}x = \alpha$. This region comprises those values of $x$ which have the largest values of $P_A(x)/P_H(x)$.

Suppose a small region of width $\Delta_1$ at $x = x_1$ is moved from the rejection to the acceptance region. It must be replaced by an equivalent region from, say, $x = x_2$, with width given by $\Delta_2 P_H(x_2) = \Delta_1 P_H(x_1)$, so that $\alpha$ stays the same. This gives an increase in $1-\beta$ if $\Delta_2 P_A(x_2) > \Delta_1 P_A(x_1)$, but, by construction, this is not the case.

To use this, you fix some value $c$, and the rejection region comprises those values of $x$ for which

$$\frac{P_A(x)}{P_H(x)} > c. \tag{8.1}$$

Any particular $c$ determines the region, and thus the values of $\alpha$ and $\beta$. If you choose to work at a particular significance level $\alpha$ this prescribes the desired value of $c$.

Perhaps the most important point about such a test is that it *only* exists for simple, completely specified, alternative hypotheses, and such cases are scarce. Usually the way to proceed is not so clear-cut.

*Example Two forms of silicon dioxide*
The density of opal is 2.2 g/cc. For quartz it is 2.6 g/cc. Various sites produce small quantities of crystals which could be either: their density is measurable with a resolution of 0.2 g/cc. Which are worth the trouble and expense of further excavation?

Hypothesize that a sample is opal. The probability of a particular density measurement $x$ is completely specified as being Gaussian of $\mu = 2.2$ and $\sigma = 0.2$, so this is a simple hypothesis. The alternative (quartz) hypothesis is also simple. The ratio of the two Gaussian is

$$\frac{\exp -(x-2.6)^2/2\sigma^2}{\exp -(x-2.2)^2/2\sigma^2} \propto \exp 10x$$

i.e. the ratio of equation 8.1 rises with $x$, and shows that any simple cut on $x$ will give optimal $\beta$ for its $\alpha$. If we accept only those samples with density $\leqslant 2.53$ ($1.64\sigma$ above the mean), then $\alpha = 5\%$, i.e. we will ignore 5% of the genuine samples. On the other hand, $\beta$ is 36%, so we will needlessly investigate 36% of the quartz deposits. The cut can be adjusted as desired, depending on the importance you attach to type I and type II errors; for example, lowering the critical value to 2.46 increases $\alpha$ to 10% and reduces $\beta$ to 24%.

## 8.2 INTERPRETING EXPERIMENTS

R.A. Fisher, one of the founding fathers of statistics, stated that 'Every experiment may be said to exist only in order to give the facts a chance of disproving the null hypothesis.' This is not quite fair of Fisher. There are lots of experiments which *measure* quantities—the velocity of sound or the viscosity of treacle—and for these there are no hypotheses, null or otherwise. His words do apply to experiments which set out to discover or *investigate* effects, and it is such experiments that require 'interpretation' in the sense of this section.

### 8.2.1 The Null Hypothesis

Suppose you are investigating ESP (extrasensory perception). A subject scores 99 correct guesses out of 100, where they would expect to get only 10 by random guesswork. However, to say that the results are consistent with the existence of ESP means nothing—they are consistent with the moon being made of green cheese, come to that. What matters is that they are *not* consistent (at some significance level) with the hypothesis of pure chance, leaving the alternative hypotheses of some extrasensory effect, or conscious or unconscious fraud.

Alice discovered in Looking-Glass country that in order to get where she wanted she had to set out in the opposite direction. Likewise in statistics, if you want to show the existence of anything, you have to hypothesize the opposite, i.e. that there is no such effect. This is called the *null hypothesis* and written $H_0$. It is an Aunt Sally you set up in order to try and knock it down again. You test it, at some level of significance, and if this test *fails* you have supported your primary aim, as the alternative to the null hypothesis is that there *is* some effect there—at what level one is not required to say.

So to show that a higher proportion of molybdenum increases the tensile strength of an alloy, or that people taking large doses of vitamin C recover from colds more quickly, or that the events in a bubble chamber are due to a new species of elementary particle, you have to hypothesize the opposite: that increased molybdenum concentration makes the alloy no stronger, that vitamin C does nothing for recovery, that the events are due to background from known processes. Paradoxically, to try and show something is there, you have to try, and fail, to assert the opposite. In statistics one cannot meaningfully accept a hypothesis: one can only reject them.

If the test of $H_0$ succeeds—say your subject had only guessed 11 right out of 100—you fail to prove the null hypothesis is false, but this does not prove that it is true. It could be that the effect is there, but at some level too small to be revealed by your experiment. You can never show that there is no effect—the best you can do is set a limit at some confidence level, as discussed in Chapter 7.

*Example Coin bias (two tailed)*

You suspect that a coin is biased, and toss it 15 times to investigate. Suppose you decide to work at the 10% significance level. Your hypothesis is that $p \neq \frac{1}{2}$, so the null hypothesis is that it is unbiased, and $p = \frac{1}{2}$. Using this assumption, some useful binomial probabilities are

$$P(15) = 0.003\% \quad P(14) = 0.05\% \quad P(13) = 0.3\%$$
$$P(12) = 1.4\% \quad P(11) = 4.2\% \quad P(10) = 9.2\%$$

where the argument of $P$ is the number of heads or, by symmetry, the number of tails. The probability of 12 or more heads is thus 1.8%; for 12 or more of the same type it is 3.5%. For 11 or more it is 11.8%, which is too much for our desired 10% level.

So if the results of 15 tosses give 12 or more of the same type, we would declare the coin to show bias, saying it was unlikely that so many could arise by chance. If they give 11 or less we would say there was no evidence for any bias in the coin.

Notice the need for the extended definition of $\alpha$ to cope with the discrete variable. There is not a number of heads/tails which gives a probability of exactly 10%, so you have to make do with the largest value below the desired limit.

*Example Coin bias (one tailed)*
Consider a coin tossing problem like the one above, except that this time your suspicion is that the coin is biased in favour of heads. Your hypothesis is now that $p > \frac{1}{2}$, so the null hypothesis is that $p \leqslant \frac{1}{2}$. The extended definition of significance now allows for the inequality arising from the indeterminacy of $H_0$. The probability under $H_0$ of 15 heads is now 0.003% or less, and the probability of 12 or more heads is 1.8% or less. We can say nothing useful about the probability of many tails—the probability of 15 tails is 0.003% or more. Setting the rejection region to cover 11 or more heads gives a $\leqslant 5.9\%$ probability. (Increase this to 10 gives $\leqslant 15.1\%$.)

So for our desired 10% level we reject the null hypothesis and declare the coin biased if there are 11 or more heads from 15 throws.

The reason for the name 'significance' is now clear: a result is 'significant' if the probability that it could have arisen by chance from the null hypothesis is small. Ideally one first decides the significance level one wants to work at, and then performs the experiment. In practice one usually performs the experiment, examines the result, and then reports its significance.

### ★8.2.2 Binomial Probabilities

Many experiments take the form of binomial tests, where you want to know whether the outcomes were affected by some particular condition. Medical trials are typical of these, and will serve to provide examples. (In real life elaborate precautions are used to eliminate bias. These are not discussed here.)

*Example Is the treatment effective?*
A new medical treatment is to be tested on 100 patients. How will you decide whether it is effective, given that 60% of sufferers are cured spontaneously within a week anyway?

Suppose we decide to work at the 5% significance level.

If the probability that a patient taking the treatment will be cured in a week is $P$, then the null hypothesis to be tested (that the treatment has no beneficial effect) is that $P \leqslant 0.6$. Under this null hypothesis we expect on average 60 cures (or less), with

$\sigma$, from the binomial of (at most)

$$\sqrt{(100 \times 0.4 \times 0.6)} = \sqrt{24} = 4.90.$$

Numbers are large, so use the normal approximation: a one-tailed test achieves 5% significance at $1.64\sigma$, which is $60 + 1.64 \times 4.9 = 68.03$ cures. So 69 cures would be needed to show that the treatment is beneficial.

Where did this number of 100 patients come from? How many patients should we use to test a new treatment? This depends on what we want. Suppose that we are only interested in a cure which is 100% effective. Then of $N$ patients all $N$ will be cured. The probability of this happening under the null hypotheses is $0.6^N$, so for a 5% significance we require that

$$0.6^N \leqq 0.05$$

which gives $N = 6$. We conduct the trial on six patients and only deem the treatment effective if all of them are cured.

More realistically, one might well want to know if a treatment increased the recovery probability to, say, 70% or more. As there is still a possibility that a 70% effective treatment may cure (much) less than 70% of the patients, we cannot now insist on identifying such a cure; we can only ask for a test with high power. Take 99% as an example.

Using the normal approximation again, we will cut as before at a number of cures equal to $0.6N + 1.64\sqrt{0.24N}$ to get the desired 5% significance. We now impose the additional requirement that if the true cure rate is 70% (or more) then the probability of failing this cut is 1% (or less). The number of standard deviations corresponding to a 99% one-sided confidence level is 2.33. So we must satisfy

$$0.6N + 1.64\sqrt{0.6 \times 0.4 \times N} \leqslant 0.7N - 2.33\sqrt{0.7 \times 0.3 \times N}$$

which gives $N \geqslant 351$. The cut, as given by both sides of this equation, falls at 226 cures.

To summarise: if 351 patients are tested, and the treatment deemed effective if 226 or more are cured, then the probability of a 70% effective treatment slipping through the net is 1%. The odds for a more effective treatment are even better. The probability of a false alarm is not more than 5%.

*Then felt I like some watcher of the skies*
*When a new planet swims into his ken*

*—Keats*

### ★8.2.3 Is There a Signal??—Poisson Statistics

Many experiments set out to discover new entities, ranging in size from elementary particles to clusters of galaxies. Many others make such discoveries serendipitously, having set out to investigate something completely different.

Such objects may first appear as an unexpected peak: a histogram of some quantity shows a high number of events concentrated at a particular point. More commonly the situation is not so clearcut: a peak is seen, but there is also a non-negligible background in that region of the histogram, and you need to take into account the possibility that your peak is not an exciting discovery, but merely a fluctuation of the unexciting background.

For example: suppose we are searching for a short-lived nuclear isotope, known to emit gamma rays of a certain energy. In the data we do indeed find 87 such events. However, there is also considerable background from the decays of other isotopes, and we would expect a background of, on average, 54 events. Do we have a real signal?

The null hypothesis is to expect 54 events. This is a Poisson process, so the error is $\sqrt{54} = 7.35$; the numbers are large enough for the normal approximation to be used. The difference is $4.5\sigma$, so the null hypothesis is rejected (unless you are amazingly cautious): there must be a signal there.

A harder version of this arises if the calculation which gives the figure for the average background is not exact, so the average has an error of its own: $\pm 3$, say, in addition to the Poisson error describing the variation of the actual figure from the average. The correct approach now is to say: we are looking at the difference between the signal and predicted background; the variance on this difference is the sum of the two variances (combination of errors) which in this case is $54 + 3^2 = 63$. This reduces it to a $4.2\sigma$ effect.

Here we were looking in a region where we knew the signal should be. If we had just looked at the graph for regions where the numbers seemed high, this is not the same and the result not as significant. Even from a single set of data, there are many bins, many ways of binning, and also many ways of cutting and selecting the data before you plot it, so you can usually come up with some sort of effect of high 'significance' if you try long enough. One learns rapidly not to get excited too quickly about such findings.

## 8.3 GOODNESS OF FIT

'Goodness of fit' is a very important variety of hypothesis testing. As usual, you have a sample of data. You also have a function which is supposed to describe the data. The question is whether to believe this: does the function really provide an adequate description of the way the data behave? Or are there significant and incompatible differences?

To find out, you proceed as described in section 8.2.1, and introduce the null hypothesis. This says that the function really does describe the data, and any differences are mere fluctuations. To show that there are real differences between the data and the function, you have to 'give the data a chance to refute' this hypothesis by finding the probability that these fluctuations could have arisen by chance. If this probability is small then the null hypothesis is rejected, the function and the data are deemed to disagree, and the fit is not 'good'. If the probability is reasonably large, the fit is accepted.

### 8.3.1 The $\chi^2$ Test

This is far and away the most important test for goodness of fit. (Details of the $\chi^2$ distribution can be found in section 6.4.)

The data consist of a set of measurements of $x$ and $y$, where the $x$ values are exact and each $y$ is measured with error $\sigma$. There is a function $f(x)$ which claims to give the ideal value of $y$ for a given $x$. Then $\chi^2$ is

$$\chi^2 = \sum_{i=1}^{N} \frac{[y_i - f(x_i)]^2}{\sigma_i^2}. \tag{8.2}$$

If the $y$ measurement errors are correlated, then this becomes a matrix equation

$$\chi^2 = (\tilde{\mathbf{y}} - \tilde{\mathbf{f}})\mathbf{V}^{-1}(\mathbf{y} - \mathbf{f}) \tag{8.3}$$

where $\mathbf{V}$ is the covariance matrix for the $y_i$ measurements.

If the function really does describe the data, then the difference between the true value and the measurement at each point should be roughly the same size as the measurement error, so we expect a contribution of about one from each term in the sum, and $\chi^2$ to be roughly equal to $N$. If $\chi^2$ is large then it is telling you that there is something wrong with the answer. That much is obvious; the task is to quantify the statement and explain what is meant by a 'large' value of $\chi^2$.

The probability distribution for $\chi^2$ is given by (see section 6.4)

$$P(\chi^2; N) = \frac{2^{-N/2}}{\Gamma(N/2)} \chi^{N-2} e^{-\chi^2/2}. \tag{8.4}$$

To make judgements and decisions about goodness-of-fit, the relevant quantity is the integral

$$P(\chi^2; N) = \int_{\chi^2}^{\infty} P(\chi'^2; N) d\chi'^2. \tag{8.5}$$

This is called the $\chi^2$ *probability*. This is a quantity very often met with, so it is worth spelling out exactly what it means. It is the probability that a function which does genuinely describe a set of $N$ data points would give a value of $\chi^2$ as large, or larger, than the one you already have.

> Suppose you have a $\chi^2$ of 20 for 5 points. This is not impossible, just unlikely. The probability that a real $\chi^2$ distribution for $n = 5$ would give a value of 20.0 or more is $\int_{20}^{\infty} P(\chi^2; 5)\, d\chi^2 = 0.0012$. The $\chi^2$ probability is thus very small, and the alternative hypothesis, that the function and data disagree, is very plausible.

If $\chi^2$ is large, so that $P(\chi^2)$ is small, it could be that the errors are underestimated, or that $f(x)$ does not describe the data very well, and this is a 'bad fit'. The definition of a 'small' $\chi^2$ probability is a matter for your judgement.

If $\chi^2$ is suspiciously small (so that $P(\chi^2) \approx 1$), this cannot be blamed on the function—the $\chi^2$ test is strictly one-tailed. It could well mean that the errors have been overestimated, or that the data have been specially selected, or that you were just lucky.

The above tells you what to expect if you are given a set of points and a function. However, suppose the data have been used in finding the function parameters, using a least squares fit for example. Then $\chi^2$ will be smaller than expected, because you have minimised it. This can be handled very easily—and this is the amazingly nice thing about the $\chi^2$ distribution. If you

TABLE 8.1.
CRITICAL $\chi^2$ VALUES

| | $P = 10\%$ | $= 5\%$ | $= 2\%$ | $= 1\%$ |
|---|---|---|---|---|
| $n = 1$ | 2.71 | 3.84 | 5.41 | 6.63 |
| 2 | 4.61 | 5.99 | 7.82 | 9.21 |
| 3 | 6.25 | 7.82 | 9.84 | 11.34 |
| 4 | 7.78 | 9.49 | 11.67 | 13.28 |
| 5 | 9.24 | 11.07 | 13.39 | 15.09 |
| 6 | 10.64 | 12.59 | 15.03 | 16.81 |
| 7 | 12.02 | 14.07 | 16.62 | 18.47 |
| 8 | 13.36 | 15.51 | 18.17 | 20.09 |
| 9 | 14.68 | 16.92 | 19.68 | 21.67 |
| 10 | 15.99 | 18.31 | 21.16 | 23.21 |
| 11 | 17.27 | 19.68 | 22.62 | 24.72 |
| 12 | 18.55 | 21.03 | 24.05 | 26.22 |
| 13 | 19.81 | 22.36 | 25.47 | 27.69 |
| 14 | 21.06 | 23.68 | 26.87 | 29.14 |
| 15 | 22.31 | 25.00 | 28.26 | 30.58 |
| 16 | 23.54 | 26.30 | 29.63 | 32.00 |
| 17 | 24.77 | 27.59 | 31.00 | 33.41 |
| 18 | 25.99 | 28.87 | 32.35 | 34.81 |
| 19 | 27.20 | 30.14 | 33.69 | 36.19 |
| 20 | 28.41 | 31.41 | 35.02 | 37.57 |
| 21 | 29.62 | 32.67 | 36.34 | 38.93 |
| 22 | 30.81 | 33.92 | 37.66 | 40.29 |
| 23 | 32.01 | 35.17 | 38.97 | 41.64 |
| 24 | 33.20 | 36.42 | 40.27 | 42.98 |
| 25 | 34.38 | 37.65 | 41.57 | 44.31 |
| 26 | 35.56 | 38.89 | 42.86 | 45.64 |
| 27 | 36.74 | 40.11 | 44.14 | 46.96 |
| 28 | 37.92 | 41.34 | 45.42 | 48.28 |
| 29 | 39.09 | 42.56 | 46.69 | 49.59 |
| 30 | 40.26 | 43.77 | 47.96 | 50.89 |

have $N$ terms in the $\chi^2$ sum and have adjusted $m$ parameters in the function (including the overall normalisation, if appropriate), to minimise this sum, then (see section 6.4.1) the appropriate distribution is still $\chi^2$, but is $P(\chi^2; N-m)$, i.e. $N$ changes to $n = N - m$. $n$ is called the *number of degrees of freedom*. Thus if 10 points are compared to a given straight line, there are 10 degrees of freedom, but if the line has been fitted through them there are only 8.

The integrated probabilities are often given on graphs, or by subroutine library calls, or by tables, such as Table 8.1.

This gives the value of $\chi^2$ for various probabilities. For example, the probability of a $\chi^2$ of 29.62 or more arising from 21 degrees of freedom is 10%. A value of 36.34 or more has a chance of only 2%.

*Example Goodness of fit using $\chi^2$*

A straight fit drawn through 20 points gives a $\chi^2$ of 36.3. A parabola gives a $\chi^2$ of 20.1, and a cubic a $\chi^2$ of 17.6.

The straight line has $20 - 2 = 18$ degrees of freedom. The table gives the probability of exceeding 34.8 as only 1%, so the probability for 36.3 is even smaller. It is most improbable that such a high value could arise by chance, and most implausible that the line really describes the data.

The $\chi^2$ for the parabola is comfortably below the 10% probability threshold (for 17 degrees of freedom); the data are well described by the parabola and there is no justification for using the more complicated cubic.

For larger values you can use the fact that $\sqrt{2\chi^2}$ follows a Gaussian distribution with mean $\sqrt{2n-1}$ and standard deviation 1. For example, suppose you want to find the 5% level for 30 degrees of freedom. The Gaussian table (3.3) gives the one-tailed 95% limit as $1.645\sigma$. The $\chi^2$ limit is accordingly given by

$$\sqrt{2\chi^2} = \sqrt{2 \times 30 - 1} + 1.645$$

which gives the value of the $\chi^2$ limit as 43.49, acceptably close to the exact value from the table of 43.77.

A common application of goodness of fit is to a histogram of a set of data values. $y_i$ is the number of events found in bin number $i$ $(i = 1, 2, 3, \ldots, n)$ which has a mean value $x_i$. $f(x_i)$ is the predicted number of events in that bin. The errors are given by Poisson statistics and $\chi^2$ is

$$\chi^2 = \sum_{i=1}^{n} \frac{[y_i - f(x_i)]^2}{f(x_i)}. \tag{8.6}$$

The number of degrees of freedom is now the number of *bins* minus the number of fitted parameters. (When counting the number of parameters, do not forget the overall normalisation.)

When binning data in this way, you have to choose suitable bin sizes. You

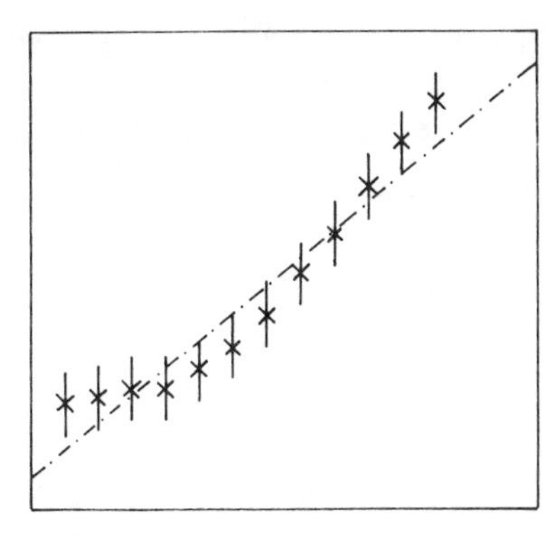

Fig. 8.3. A straight line through twelve data points.

need at least five events per bin for a meaningful $\chi^2$, and ten is better. On the other hand, if you make the bins too large this can obscure significant detail. For best results it may be necessary to use bin widths of varying sizes, such that the numbers in each bin are roughly the same. This can be done either on the basis of the known function $f(x)$, or from the data, starting with a very fine bin size and merging them till the desired balance is attained.

### ★8.3.2 The Run Test

Consider the data and fitted line shown in Figure 8.3. It is obvious that something is not right, and the data really need to be fitted by a higher power. Nevertheless, the points lie reasonably close to the curve and the $\chi^2$ is reasonable (in fact it is 12.0 for the twelve points)—possibly the errors have been overestimated. In cases like this the *run test* provides vital extra information.

There are six points below the line and six above. In order of increasing $x$ they are in the sequence $AAABBBBBBAAA$, writing $A$ and $B$ for 'above' and 'below'. They give a run of $A$s, then a run of $B$s, then another run of $A$s, making three runs in all. This seems suspiciously small—in a 'good' fit the points above and below should be jumbled together, giving many (short) runs.

The probability of the $A$s and $B$s giving a particular number of runs, $r$, can be calculated. Suppose there are $N_A$ points above the line and $N_B$ points below, with $N_A + N_B = N$. The number of different ways they can be arranged is

$$_NC_{N_A} = \frac{N!}{N_A!N_B!}.$$

Now we find how many of these correspond to a particular number of runs $r$. Suppose first that $r$ is even and that the sequence starts with an $A$. Forgetting about the $B$s for a moment, that means there are $N_A$ $A$-points, and $r/2-1$ divisions between them. For example, the twelve runs

$$AAABBABBBABAABAABBBBABB$$

would be considered as

$$AAA|A|A|AA|AA|A$$

With the $N_A$ points in sequence, there are $N_A-1$ places to put the first dividing line (it cannot go at the ends), then $N_A-2$ choices for the second (it cannot go at the ends or next to the first dividing line), and so on, giving ${}_{N_A-1}C_{r/2-1}$ different possibilities for the $A$ arrangements. There is a similar factor from arranging the $B$s, and another factor of 2 because the first point could have been a $B$. Thus the probability of $r$ runs is, for $r$ even,

$$P_r = 2\,\frac{{}_{(N_A-1)}C_{(r/2-1)} \times {}_{(N_B-1)}C_{(r/2-1)}}{{}_{N}C_{N_A}}. \tag{8.7}$$

Similarly, for $r$ odd it is

$$\frac{{}_{(N_A-1)}C_{(r-3/2)} \times {}_{(N_B-1)}C_{(r-1/2)} + {}_{(N_A-1)}C_{(r-1/2)} \times {}_{(N_B-1)}C_{(r-3/2)}}{{}_{N}C_{N_A}}. \tag{8.8}$$

From these it can be shown that

$$\langle r \rangle = 1 + \frac{2N_AN_B}{N} \tag{8.9}$$

$$V(r) = \frac{2N_AN_B(2N_AN_B-N)}{N^2(N-1)}. \tag{8.10}$$

For $N_A$, $N_B$ greater than about 10 to 15, one can use the Gaussian approximation for $r$. For smaller numbers, one can calculate the sums using equations 8.7 and 8.8, or use tables.

In our example, these formulae show that the average number of runs to expect is $\langle r \rangle = 1 + 2 \times 6 \times 6/(6+6) = 7$, with a variance of 2.73 and thus $\sigma$ of 1.65. The deviation of $7-3=4$ thus constitutes 2.4 standard deviations, which is significant at the 1% level (one-tailed test). Thus, despite the verdict of the $\chi^2$ test, the run test gives good ground for rejecting the null hypothesis and saying the fit is not good.

The run test is much less powerful than the $\chi^2$ test, but it provides *additional* useful independent information. A fit may have an acceptable $\chi^2$, perhaps because of overestimated errors, and still be wrong and rejectable by the run test. The $\chi^2$ test ignores the sign of the deviations; the run test looks only at the signs.

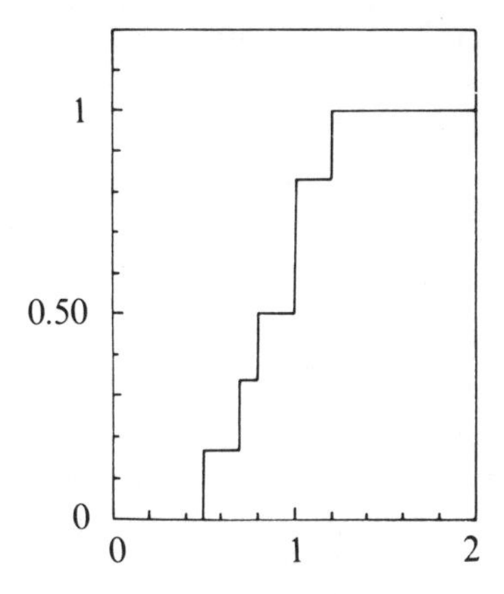


Fig. 8.4. The cumulative distribution for the set of data 0.5, 0.7, 0.8, 1.0, 1.0, 1.2.

### ★8.3.3 The Kolmogorov Test

The Kolmogorov test is an alternative to the $\chi^2$ test when the data sample is small so you cannot find a sensible bin size. It does not involve any binning so no information is 'thrown away'.

You take the values and arrange them in increasing order, and plot the *cumulative* distribution cum$(x)$, divided by the total number of values $N$. This gives a function like a staircase going from a level of 0 to 1, going up a step of $1/N$ at every data point (Figure 8.4).

You also plot the cumulative distribution cum$(P)$ for the probability distribution function $P$, which will be a smooth curve. The greatest absolute difference between the two curves, i.e. where they are furthest apart, is $D = \max|\text{cum}(x) - \text{cum}(P)|$. This is multiplied by $\sqrt{N}$ to get a value $d = D\sqrt{N}$. If the agreement is good then $d$ should be small; in quantitative terms there are critical values for $d$.

| Critical value | Significance |
|---|---|
| 1.63 | 1% |
| 1.36 | 5% |
| 1.22 | 10% |
| 1.07 | 20% |

These values only apply for large values of $N$. If it is not large, then you have to consult a more complicated table.

This test is often used to test whether a set of data is uniformly distributed over its range. The curve cum $(P)$ is then a straight line.

Unfortunately the Kolmogorov test only applies when the distribution is fixed beforehand. If it has been fitted to the data in some way, this test cannot be used—there is no equivalent of the 'degrees of freedom' treatment of the $\chi^2$ test.

## 8.4 THE TWO-SAMPLE PROBLEM

One common problem is to take two samples and ask whether they are compatible. An experimental sample and a control, chocolate sales before and after an advertising campaign, production line failure rates with and without quality control—the examples are endless.

Even if two samples are taken from same parent, they will have differences due to statistical fluctuations. You have to decide whether the differences between the two samples are explainable by such random effects, or whether they are too large for that and indicate a real difference in the nature of the two samples. The problem is a bit like goodness of fit, except that instead a comparing a sample with a distribution, you are comparing one sample with another.

Unfortunately 'are they same' is not a very precise way of specifying the problem. Furthermore, it is important to know how much can be safely assumed before we start; the more assumptions that can be made, the more powerful the test can be. In particular, if the distributions are known to be Gaussian (which, thanks to the central limit theorem, they often are) that helps a lot.

We will consider the two-sample problem in five forms:

1. Two Gaussian samples with known $\sigma$. Are the means the same?
2. Two Gaussian samples with the same (unknown) $\sigma$. Are the means the same?
3. Matched and correlated samples: how to beat $\sqrt{N}$.
4. Two Gaussian distributions. Is $\sigma$ the same?
5. The completely general problem.

### 8.4.1 Two Gaussian Samples with Known $\sigma$

There are two values, $X$ and $Y$. They are known to be randomly distributed about their true value(s) according to the Gaussian distribution with standard deviations $\sigma_X$ and $\sigma_Y$. The question is whether $X$ and $Y$ could 'really' be the same.

This is equivalent to asking whether $X - Y$ is compatible with zero. This is easy to analyse. $X - Y$ has a variance of

$$V(X - Y) = \sigma_X^2 + \sigma_Y^2. \tag{8.11}$$

The question thus reduces to how many $\sigma$ the difference is from zero, and a decision made at a particular significance level at a value determined from the table of the integrated Gaussian.

This sort of situation is very common. There are two particular instances:

1. Two experimental measurements, each with apparatus of known resolution, $\sigma_X$ and $\sigma_Y$.
2. Two samples taken under different conditions, where the variances of the parent distributions are known. The two values are then the means of the two samples, $\bar{x}$ and $\bar{y}$, and the standard deviations for these are the standard errors on the means, $\sigma_x/\sqrt{N_x}$ and $\sigma_y/\sqrt{N_y}$.

*Example Two measurements*
The melting point of two specimens of material are measured as $202 \pm 3$ K and $209 \pm 4$ K, where the errors are independent. Could these be two samples of the same substance?

The error is $\sqrt{3^2 + 4^2} = 5$ K. The difference of 7 K is $1.4\,\sigma$ and thus quite acceptable. There is no reason to believe the substances to be different.

*Example Two samples*
One year the average A-level score for students was 12.5 for 109 students. The next year it was 12.7 for 123 students. The distributions have a spread of $\sigma = 2.0$. Is the standard improving?

The null hypothesis is that there is no change between the two years. The difference is $12.7 - 12.5 = 0.2$. The errors on the two average values are $2.0/\sqrt{109} = 0.19$ and $2.0/\sqrt{123} = 0.18$ respectively. These add in quadrature to give an error of 0.26. The increase is less than one standard deviation and there is no reason for rejecting the null hypothesis—in plain English, there are no grounds for claiming an improvement.

### ★8.4.2 Two Gaussian Samples with Unknown $\sigma$

Suppose you want to compare two values, known to be produced by Gaussian distributions, but with no knowledge of the standard deviations. If these are two bare measurements (as in case (1) of the previous section) then there is nothing you can do, but if they are the averages of two samples, the spread of the values within the samples can be used to estimate $\sigma$:

$$\hat{\sigma}_x = s_x = \sqrt{\frac{\sum(x_i - \bar{x})^2}{N_x - 1}} \qquad \hat{\sigma}_y = s_y = \sqrt{\frac{\sum(y_i - \bar{y})^2}{N_y - 1}}. \tag{8.12}$$

Because we are using an estimate of $\sigma$, rather than the true value, we have to use Student's $t$ to test the significance instead of the Gaussian (see section

7.3 for details of Student's $t$). This requires us to test a variable which is made from a Gaussian distribution divided by a $\chi^2$. Now, under the null hypothesis, which is $\mu_1 = \mu_2$, the difference $(\bar{x} - \bar{y})/\sqrt{(\sigma_x^2/N_x) + (\sigma_y^2/N_y)}$ is Gaussian of mean zero and standard deviation unity, which provides the numerator. The sum

$$\frac{(N_x - 1)s_x^2}{\sigma_x^2} + \frac{(N_y - 1)s_y^2}{\sigma_y^2}$$

is a $\chi^2$ with $N_x + N_y - 2$ degrees of freedom, as can be seen by inserting equation 8.12. Two degrees of freedom are lost, as the two means are unknown.

You can form a $t$ variable from the ratio. To be useful the $\sigma$ factors have to cancel when this is done, and this happens if $\sigma_1 = \sigma_2$. So the variable

$$\frac{\bar{x} - \bar{y}}{S\sqrt{(1/N_x) + (1/N_y)}} \tag{8.13}$$

where

$$S^2 = \frac{(N_x - 1)s_x^2 + (N_y - 1)s_y^2}{N_x + N_y - 2} \tag{8.14}$$

is distributed according to Student's $t$ distribution, with the number of degrees of freedom $n = N_x + N_y - 2$, and can be tested accordingly.

$S$ is called the 'pooled estimate' of the standard deviation, as it is the combined estimate from the two samples, appropriately weighted. The term $S\sqrt{(1/N_x) + (1/N_y)}$ is analogous to the standard error on the mean $\sigma/\sqrt{N}$ that is used when $\sigma$ is known.

*Example Student's t for two samples*

The marks (out of 50) scored in a maths exam by a group of physics students and a group of engineering students are as follows (note that for the sake of harmony, the two groups are here denoted 1 and 2, so you cannot tell which is which):

| Group 1 | 39 | 18 | 3 | 22 | 24 | 29 | 22 | 22 | 27 | 28 | 23 | 48 |
|---|---|---|---|---|---|---|---|---|---|---|---|---|
| Group 2 | 42 | 23 | 36 | 35 | 38 | 42 | 33 | | | | | |

Group 1 has twelve students, a mean $\bar{x}_1$ of 25.4 marks, and an estimated $\sigma$ of $s_1 = 10.9$. Group 2 contains seven students, with $\bar{x}_2$ of 35.6, and $s_2 = 6.5$. The numerator in equation 8.13 is thus 10.2, and equation 8.14 gives the pooled estimate $S$ as 9.6 (which seems reasonable—it is between $s_1$ and $s_2$ and closer to $s_1$, as that is the larger sample). This is multiplied by $\sqrt{1/7 + 1/12} = 0.48$ to give the denominator of equation 8.13 as 4.6. $t$ is thus $10.2/4.6 = 2.2$. Consulting Table 7.2, for 17 degrees of freedom, using the two-tailed values, this exceeds the 95% confidence limit $t$ value of 2.110 (though not the 98% limit). We would therefore say that the groups are different (at the 5% significance level).

You might be worried in the example just given, as we have not given reasons for assuming that exam marks follow a Gaussian distribution. The central limit theorem can be appealed to, and gives some support, but not a full guarantee: performance depends on several factors, but not an infinite number. In practice one expects the distribution to have a bump in the middle and tail off to zero at high and low values, but there is no reason to expect it to be a perfect Gaussian shape—it will very probably have a negative skew due to a tail of students with poor marks.

Most users of the test do not bother about the Gaussian assumption; they just plug the numbers into the formula and produce an answer. And they get away with it. In fact the test of Student's $t$ works remarkably well, even for distributions that are not accurately Gaussian. It is *robust*, to use the technical term.

Our other unjustified assumption is that the standard deviations of the two groups are the same. This can be properly analysed, and this is done in section 8.4.4.

Nevertheless, no matter how complicated or convincing the significance tests used, the conclusions that can be drawn from analyses like this depend entirely on the assumptions that are made. If the quoted results is to mean anything these have to be honestly specified.

### ★8.4.3 Matched and Correlated Samples

*Matched* or *correlated* samples are another way of beating $\sqrt{N}$, similar to the method of stratified sampling (see section 5.7).

Given two sets of data, with the same number of measurements in each, we want to know if the mean values are compatible. The previous approach (sections 8.4.1 and 8.4.2) has been to find the difference of two means, $\bar{x}$ and $\bar{y}$, and compare that with zero, using tables of integrated Gaussian if the standard deviations are known, or Student's $t$ if they are not.

Now, the individual means will vary for all sorts of reasons, many of which are of no interest. We can reduce this effect by correlating the members of the samples—by comparing like with like. An extreme case is the 'before and after' matching, where you compare each member with itself, before and after some sort of treatment.

In terms of the obvious mathematical relation

$$\sum x_i - \sum y_i = \sum (x_i - y_i)$$

we are examining the right-hand side rather than the left. This in itself does nothing, unless the measurements are paired together in a meaningful way. The distribution in $x_i - y_i$ has variance

$$\sigma_x^2 + \sigma_y^2 - 2\rho\sigma_x\sigma_y$$

so that if the samples are correlated, the variance is smaller.

To use Gaussian tables, as in section 8.4.1, you now have to know the correlation, as well as $\sigma_x$ and $\sigma_y$. For the Student's $t$ analysis of section 8.4.2 you do not need to know anything about it; you just take the values of $x_i - y_i$ and use their actual spread.

*Example Matched pairs*
A consumer magazine is testing a widget claimed to increase fuel economy. Here are the data on seven cars. Is there evidence for any improvement?

| Car | A | B | C | D | E | F | G |
|---|---|---|---|---|---|---|---|
| Miles per gallon without widget | 29 | 30 | 42 | 34 | 37 | 45 | 32 |
| Miles per gallon with widget | 36 | 26 | 46 | 36 | 40 | 51 | 35 |
| Difference | 7 | −4 | 4 | 2 | 3 | 6 | 3 |

If you ignore the matching, the means are $38.6 \pm 3.1$ and $35.6 \pm 2.3$ for the samples with and without the widget. The improvement of 3 m.p.g. is not significant as it is smaller than the errors.

Now look at the differences. Their average is 3.0. The estimated standard deviation $s$ is 3.6, so the error on the estimated average is $3.6/\sqrt{7} = 1.3$, and $t$ is $3.6/1.3 = 2.0$. This is significant at the 5% level using Student's $t$ (one-tailed test, 6 degrees of freedom, $t_{\text{crit}} = 1.943$).

### ★8.4.4 The $F$ Distribution

Sometimes you want to know whether the variances of two samples are the same, either out of interest in their own right or to find out whether you can validly apply Student's $t$ as in section 8.4.2 (for which the variances had to be the same).

To do this you take the estimated variances of the two samples and form the ratio

$$F = \frac{\hat{V}_1}{\hat{V}_2} \tag{8.15}$$

where (as introduced in equation 5.14) $\hat{V} = \sum (x_i - \bar{x})^2/(N-1)$ for either sample. The ratio will be near 1 if the two parent variances are indeed the same, and a significant difference from 1 means the variances are different.

More specifically: when divided by $\sigma^2$ the variance is distributed according to a $\chi^2$ distribution with $N - 1$ degrees of freedom. Because we have formed the ratio, the factors of $\sigma^2$ cancel, and the variable $F$ consists of the ratio of two $\chi^2$ distributions, normalized by their number of degrees of freedom: $f_1 = N_1 - 1$ in the top and $f_2 = N_2 - 1$ degrees of freedom in the bottom.

The distribution for such a quantity can be shown to be

$$P(F) = \frac{\Gamma((f_1 + f_2)/2)}{\Gamma(f_1/2)\Gamma(f_2/2)} \sqrt{f_1^{f_1} f_2^{f_2}} \frac{F^{(f_1/2)-1}}{(f_2 + f_1 F)^{(f_1+f_2)/2}}.$$

Names are a confusion again here. This is known as the *F distribution*, the *Snedecor distribution*, the *Fisher distribution*, and the *variance ratio distribution*. It is even more complicated than Students $t$, as there are now two degrees of freedom to worry about.

As it does not matter which way round you form the ratio, the convention is to put the larger value on top, whichever that may be, so that the $F$ you get is always bigger than 1. A very large $F$ indicates that the variances are different, and $F$ close to 1 means they can be taken as the same, where 'large' and 'close' need to be defined taking into account $f_1, f_2$, and the desired confidence level. For large numbers

$$Z = \tfrac{1}{2} \log F$$

has a distribution that is a reasonable approximation to the Gaussian, with mean $\frac{1}{2}(1/f_2 - 1/f_1)$ and variance $\frac{1}{2}(1/f_2 + 1/f_1)$.

If the numbers of degrees of freedom are smaller, tables must be used. Table 8.2 and 8.3 give two examples—extensive tables exist for serious users.

The column refers to the number of degrees of freedom in the numerator, which is chosen as the larger of the two variances. The row refers to the denominator. For example, if and only if the greater variance (with 4 degrees of freedom) exceeds the lesser (with 5) by more than a factor of 3.52 is the difference significant (at the 10% level).

*Example Two variances*
In the example in section 8.4.2 we found two samples, one with twelve members and a variance of 10.9, another with seven members and a variance of 6.5. The ratio $F$ is 1.68. Consulting the 10% significance table, for row 11 and column 6, the critical value is given as 2.92, and our figure is safely below this. So there is no statistical evidence for any difference in the variances, and the use of Student's $t$ is compatible with the data given.

### ★8.4.5 The General Case

The most general form of the two-sample problem makes no assumptions at all about the distribution, and asks not about the mean or the variance but whether 'the distributions the same'.

The run test (as described in section 8.3.2) can be used for this purpose. It works as follows. Arrange both samples *together* in order, to get a sequence of $x$'s and $y$'s.

If the $x$ and $y$ values are drawn from the same distribution, they will be

TABLE 8.2
*F* DISTRIBUTION CRITICAL VALUES FOR 5% SIGNIFICANCE

| | 1 | 2 | 3 | 4 | 5 | 6 | 7 | 8 | 9 | 10 | 11 | 15 | 20 | 50 | 100 |
|---|---|---|---|---|---|---|---|---|---|---|---|---|---|---|---|
| 1 | 161 | 199 | 216 | 225 | 230 | 234 | 237 | 239 | 241 | 242 | 243 | 246 | 248 | 252 | 253 |
| 2 | 18.5 | 19.0 | 19.2 | 19.2 | 19.3 | 19.3 | 19.4 | 19.4 | 19.4 | 19.4 | 19.4 | 19.4 | 19.4 | 19.5 | 19.5 |
| 3 | 10.1 | 9.55 | 9.28 | 9.12 | 9.01 | 8.94 | 8.69 | 8.85 | 8.81 | 8.79 | 8.76 | 8.70 | 8.66 | 8.58 | 8.54 |
| 4 | 7.71 | 6.94 | 6.59 | 6.39 | 6.26 | 6.16 | 6.09 | 6.04 | 6.00 | 5.96 | 5.94 | 5.86 | 5.80 | 5.70 | 5.66 |
| 5 | 6.61 | 5.79 | 5.41 | 5.19 | 5.05 | 4.95 | 4.88 | 4.82 | 4.77 | 4.74 | 4.70 | 4.62 | 4.56 | 4.44 | 4.40 |
| 6 | 5.99 | 5.14 | 4.76 | 4.53 | 4.39 | 4.28 | 4.21 | 4.15 | 4.10 | 4.06 | 4.03 | 3.94 | 3.87 | 3.75 | 3.71 |
| 7 | 5.59 | 4.74 | 4.35 | 4.12 | 3.97 | 3.87 | 3.79 | 3.73 | 3.68 | 3.64 | 3.60 | 3.51 | 3.44 | 3.32 | 3.27 |
| 8 | 5.32 | 4.46 | 4.07 | 3.84 | 3.69 | 3.58 | 3.50 | 3.44 | 3.39 | 3.35 | 3.31 | 3.22 | 3.15 | 3.02 | 2.97 |
| 9 | 5.12 | 4.26 | 3.86 | 3.63 | 3.48 | 3.37 | 3.29 | 3.23 | 3.18 | 3.14 | 3.10 | 3.01 | 2.94 | 2.80 | 2.76 |
| 10 | 4.96 | 4.10 | 3.71 | 3.48 | 3.33 | 3.22 | 3.14 | 3.07 | 3.02 | 2.98 | 2.94 | 2.84 | 2.77 | 2.64 | 2.59 |
| 11 | 4.84 | 3.98 | 3.59 | 3.36 | 3.20 | 3.09 | 3.01 | 2.95 | 2.90 | 2.85 | 2.82 | 2.72 | 2.65 | 2.51 | 2.46 |
| 15 | 4.54 | 3.68 | 3.29 | 3.06 | 2.90 | 2.79 | 2.71 | 2.64 | 2.59 | 2.54 | 2.51 | 2.40 | 2.33 | 2.18 | 2.12 |
| 20 | 4.35 | 3.49 | 3.10 | 2.87 | 2.71 | 2.60 | 2.51 | 2.45 | 2.39 | 2.35 | 2.31 | 2.20 | 2.12 | 1.97 | 1.91 |
| 50 | 4.03 | 3.18 | 2.79 | 2.56 | 2.40 | 2.29 | 2.20 | 2.13 | 2.07 | 2.03 | 1.99 | 1.87 | 1.78 | 1.60 | 1.52 |
| 100 | 3.94 | 3.09 | 2.70 | 2.46 | 2.31 | 2.19 | 2.10 | 2.03 | 1.97 | 1.93 | 1.89 | 1.77 | 1.68 | 1.48 | 1.39 |

TABLE 8.3

$F$ DISTRIBUTION CRITICAL VALUES FOR 10% SIGNIFICANCE

| | 1 | 2 | 3 | 4 | 5 | 6 | 7 | 8 | 9 | 10 | 11 | 15 | 20 | 50 | 100 |
|---|---|---|---|---|---|---|---|---|---|---|---|---|---|---|---|
| 1 | 39.9 | 49.5 | 53.6 | 55.8 | 57.2 | 58.2 | 59.1 | 59.7 | 60.1 | 60.5 | 60.7 | 61.5 | 62.0 | 63.0 | 63.3 |
| 2 | 8.53 | 9.00 | 9.16 | 9.24 | 9.29 | 9.33 | 9.35 | 9.37 | 9.38 | 9.39 | 9.40 | 9.43 | 9.44 | 9.47 | 9.48 |
| 3 | 5.54 | 5.46 | 5.39 | 5.34 | 5.31 | 5.28 | 5.27 | 5.25 | 5.24 | 5.23 | 5.22 | 5.20 | 5.18 | 5.15 | 5.14 |
| 4 | 4.54 | 4.32 | 4.19 | 4.11 | 4.05 | 4.01 | 3.98 | 3.95 | 3.94 | 3.92 | 3.91 | 3.87 | 3.84 | 3.79 | 3.78 |
| 5 | 4.06 | 3.78 | 3.62 | 3.52 | 3.45 | 3.40 | 3.37 | 3.34 | 3.32 | 3.30 | 3.28 | 3.24 | 3.21 | 3.15 | 3.13 |
| 6 | 3.78 | 3.46 | 3.29 | 3.18 | 3.11 | 3.05 | 3.01 | 2.98 | 2.96 | 2.94 | 2.92 | 2.87 | 2.84 | 2.77 | 2.75 |
| 7 | 3.59 | 3.26 | 3.07 | 2.96 | 2.88 | 2.83 | 2.78 | 2.75 | 2.72 | 2.70 | 2.68 | 2.63 | 2.59 | 2.52 | 2.50 |
| 8 | 3.46 | 3.11 | 2.92 | 2.81 | 2.73 | 2.67 | 2.62 | 2.59 | 2.56 | 2.54 | 2.52 | 2.46 | 2.42 | 2.35 | 2.32 |
| 9 | 3.36 | 3.01 | 2.81 | 2.69 | 2.61 | 2.55 | 2.51 | 2.47 | 2.44 | 2.42 | 2.40 | 2.34 | 2.30 | 2.22 | 2.19 |
| 10 | 3.29 | 2.92 | 2.73 | 2.61 | 2.52 | 2.46 | 2.41 | 2.38 | 2.35 | 2.32 | 2.30 | 2.24 | 2.20 | 2.12 | 2.09 |
| 11 | 3.23 | 2.86 | 2.66 | 2.54 | 2.45 | 2.39 | 2.34 | 2.30 | 2.27 | 2.25 | 2.23 | 2.17 | 2.12 | 2.04 | 2.01 |
| 15 | 3.07 | 2.70 | 2.49 | 2.36 | 2.27 | 2.21 | 2.16 | 2.12 | 2.09 | 2.06 | 2.04 | 1.97 | 1.92 | 1.83 | 1.79 |
| 20 | 2.97 | 2.59 | 2.38 | 2.25 | 2.16 | 2.09 | 2.04 | 2.00 | 1.96 | 1.94 | 1.91 | 1.84 | 1.79 | 1.69 | 1.65 |
| 50 | 2.81 | 2.41 | 2.20 | 2.06 | 1.97 | 1.90 | 1.84 | 1.80 | 1.76 | 1.73 | 1.70 | 1.63 | 1.57 | 1.44 | 1.39 |
| 100 | 2.76 | 2.36 | 2.14 | 2.00 | 1.91 | 1.83 | 1.78 | 1.73 | 1.69 | 1.66 | 1.64 | 1.56 | 1.49 | 1.35 | 1.29 |

all mixed up, the runs of consecutive values from the same sample will be fairly short, and there will thus be a lot of them. On the other hand, if the $x$ and $y$ distributions are different, the runs will be long, and there will be only a few of them.

To consider two extreme cases: if the averages of the $x$ and $y$ distributions are very different, so that all the $x$'s come before the $y$,

$$xxxxxxxxxxxxxxxxxyyyyyyyyyyyyyyyyyy$$

then there are only two runs. If the averages are similar, but the variance of $x$ is much smaller than that of $y$, then the sequence has only three runs:

$$yyyyyyyyyyyyyxxxxxxxxxxxxxxxxyyyyyyyyyyyyyy$$

The same numerical argument apply as in section 8.3.2, and the number of runs can be tested. This test is useful only if $N_x \approx N_y$, for if they are different you get long runs anyway. Even under favourable circumstances it is not very powerful. This is not surprising as it is *completely* general—no assumptions are made about the form of the distributions, and only the order of the $x$ and $y$ data is used, not the actual numerical values.

The Kolmogorov test can also be used as a completely general test for the two-sample problem, similar to its use in goodness of fit (see section 8.3.3). In this case you plot the two cumulative distributions for $x$ and $y$ separately, divided by the appropriate total number, to get two staircases going from a level of 0 to 1, going up a step of $1/N_x$ or $1/N_y$ as appropriate at every data point.

Then find the greatest absolute difference between the two distributions and call this $D = \max|\mathrm{cum}(x) - \mathrm{cum}(y)|$. This is scaled to give $d = \sqrt{N_x N_y/(N_x + N_y)}D$ and if it is too large for your desired significance level the samples must be different. The critical values are the same as given in section 8.3.3.

## ★8.5 ANALYSIS METHODS FOR SEVERAL SAMPLES

### ★8.5.1 The Analysis of Variance (Basic Method)

For workers in the social sciences, studying similar features of several different groups, the *analysis of variance*—often abbreviated to ANOVA—is a vital and much used technique, and a great deal of literature is devoted to it. Because so much detail is readily available elsewhere, and because its use to physical scientists is more limited, only a brief outline will be given here, enough to enable you to get to grips with the specialist literature should you ever need to do so.

The samples are assumed to be Gaussian, of the same (unknown) variance, and the question is whether their means are compatible.

The $N$ elements of data are divided into $n$ groups or samples, with group $g$ having $N_g$ members, mean $\bar{x}_g$, and measured variance $V_g$. The complete data have a total overall mean $\bar{x}$ and measured variance $V$. Each group has presumably a 'true' mean $\mu_g$, of which $\bar{x}_g$ is an estimate, and there is also a true total mean $\mu$. The null hypothesis is that all the samples are compatible, so that all the $\mu_g$ are equal to each other and to $\mu$. The alternative hypothesis is that there is some real difference between the $\mu_g$.

Differences in the $\mu_g$ will cause the $\bar{x}_g$ to differ from the total average $\bar{x}$, but such differences will arise anyway due to statistical fluctuations. The problem is to decide whether the spread of the $\bar{x}_g$ about $\bar{x}$ is small enough to be compatible with the null hypothesis.

Under the null hypothesis this spread of the $\bar{x}_g$ *between groups* comes from the random nature of the measurements due to the (true) standard deviation $\sigma$. Unfortunately the value of $\sigma$ is not known. However, it can be estimated from the data itself, looking at the variation of the data within groups. The analysis therefore consists of estimating the variance of the measurements in two ways, between groups and within groups. These are tested using the Snedecor $F$ distribution described in section 8.4.4, and if the between-group variance estimate is meaningfully greater than the within-group estimate, this gives grounds for rejecting the null hypothesis and claiming evidence for variation between the samples.

The numerator used in the $F$ distribution test, the between-group variance estimate, follows from the fact that the expected error for $\bar{x}_g$ is $\sigma/\sqrt{N_g}$, so that

$$\sum_g \frac{(\bar{x}_g - \bar{x})^2}{\sigma^2/N_g} = \sum_g \frac{N_g(\bar{x}_g - \bar{x})^2}{\sigma^2}$$

is distributed like a $\chi^2$ and

$$\frac{1}{n-1}\sum_g N_g(\bar{x}_g - \bar{x})^2 \tag{8.16}$$

is the appropriate numerator for $F$. There are $n-1$ degrees of freedom as $\bar{x}$ is given by the $\bar{x}_g$.

The denominator is the estimate of $\sigma$ pooled between the groups, is in equation 8.14:

$$\frac{1}{N-n}\sum_g \sum_{i \in g} (x_i - \bar{x}_g)^2 \tag{8.17}$$

(where $g$ denotes the group and $i$ an element within that group). When divided by $\sigma^2$ the double sum is distributed like $\chi^2$ with $N-n$ degrees of freedom.

The ratio is taken and compared with the critical value at the desired significance level. This can be done for any number of groups, although if there are only two, the treatment is formally identical to the analysis using Student's $t$ described in section 8.4.2, and it is only useful when the number of groups is three or more.

*Example Behaviour of share movements*
Here are the movements of a small sample of shares in various sectors on a particular day. Is there any difference in the behaviour of different sectors?

| | | | | | | | | | | | | |
|---|---|---|---|---|---|---|---|---|---|---|---|---|
| Industrials | 0 | +1 | +1 | −1 | +2 | −1 | 0 | −4 | +4 | −1 | +2 | 0 |
| Financials | +3 | +5 | +1 | +3 | 0 | 0 | +1 | −1 | 0 | −7 | −3 | |
| Textiles | 0 | −2 | +8 | +3 | +7 | −7 | | | | | | |

The total average $\bar{x}$ is an increase of 0.48.
The average for the 12 industrials is 0.25, differing from $\bar{x}$ by 0.23.
The average for the 11 financials is 0.18, differing from $\bar{x}$ by 0.30.
The average for the 6 textiles is 1.50, differing from $\bar{x}$ by 1.02.
Equation 8.16 thus gives the numerator for the variance-ratio test as

$$\tfrac{1}{2}(12 \times 0.23^2 + 11 \times 0.30^2 + 6 \times 1.02^2) = 3.93.$$

The within-group variances are 44.3, 103.6, and 161.5. There are $12 + 11 + 6 = 29$ values and thus $29 - 3 = 26$ degrees of freedom, so the denominator term is $(44.3 + 103.6 + 161.5)/26 = 11.9$.

The numerator term is actually less than the denominator term, so there is no evidence that the variation between groups is greater than can be accounted for by the variation within groups. The performance of shares of different types was compatible; there is no cause to suppose that shares of one type did better or worse than those of another.

If the numerator had been greater than denominator, then the table of critical values for $F$ would have had to be used.

### ★8.5.2 Multiway Analysis of Variance

Another way to look at the analysis of variance described in the preceding section is as follows. The complete data have a variance (forgetting the normalisation)

$$V = \sum_{i=\text{all values}} (x_i - \bar{x})^2.$$

If the values are classified into groups, this can be written

$$\begin{aligned} V &= \sum_g \sum_{i\epsilon g} (x_i - \bar{x})^2 \\ &= \sum_g \sum_{i\epsilon g} (x_i - \bar{x}_g + \bar{x}_g - \bar{x})^2 \\ &= \sum_g \sum_{i\epsilon g} (x_i - \bar{x}_g)^2 + (\bar{x}_g - \bar{x})^2 \\ &= \sum_g V_g + \sum_g N_g(\bar{x}_g - \bar{x})^2. \end{aligned} \tag{8.18}$$

The overall variance of the data has been analysed and shown to be due to a residual variation $\sum_g V_g$ of the points themselves and a variation $\sum_g N_g(\bar{x}_g - \bar{x})^2$ between the different groups. The $N-1$ degrees of freedom are similarly partitioned into $N-n$ for the former and $n-1$ for the latter.

These two variances and numbers of degrees of freedom give the numerator and denominator for the $F$ distribution tests, but let us leave that aside. Suppose the measurements can be classified not merely by one index $g$ but by two. For example, measurements may be done by several experimenters using several different sets of apparatus, sales figures can be classified by district and by salesman, crop yields may be given for different types of crop and different fertiliser treatments. Guided by the central limit theorem, which says that variances from different effects add, the variation between the values is ascribable partly to differences in one of the two factors, partly to differences in the other, and partly to an innate residual variation . We would like to be able to partition the total variances between these three sources in the same way as it is partitioned into two in equation 8.18.

This can be done, with one important proviso: for each possible pair of values of the two factors there must be one measurement value. This means that the data can be displayed in a table, so we will refer to two factors as the *row index* $r$ and *column index* $c$.

*Example Two factors*
Twelve measurements of the velocity of light in air (in units of $10^8$ m/s), performed by four students using three different sets of apparatus, all with the same resolution, can be written as

| | Student 1 | Student 2 | Student 3 | Student 4 |
|---|---|---|---|---|
| Apparatus 1 | 3.01 | 3.03 | 3.00 | 3.04 |
| Apparatus 2 | 2.93 | 2.96 | 2.95 | 3.00 |
| Apparatus 3 | 3.00 | 3.01 | 2.99 | 3.02 |

On the face of it, we might suspect that as well as the inevitable measurement errors, there is some systematic variation between the students (student 4 seems to get high values) and also between sets of apparatus (apparatus 2 seems to give low values). We want to investigate this.

Tackling the rows first, we can separate $V$ into a within-rows variance and a between-rows variance as before:

$$V = \sum_r \sum_c (x_{rc} - \bar{x})^2$$

$$= \sum_r \sum_c (x_{rc} - \bar{x}_r)^2 + n_c \sum_r (\bar{x}_r - \bar{x})^2 \qquad (8.19)$$

where $\bar{x}_r$ is the mean for row $r$. Comparing with equation 8.18, the only point to notice is that $n_c$, the number of columns, is equal to the number of elements within each row. It is the same for all rows and so it can be taken outside the summation sign.

The first term contains the effect of the variance between the columns and any residual variance. Reversing the order of summation, it can be written as

$$\sum_c \sum_r (x_{rc} - \bar{x}_r)^2 \tag{8.20}$$

which can be separated into the variation between columns and the variation within columns. As the total sum is the same however you write it (thanks to the fact that each pair of values appears once),

$$\sum_c \sum_r x_{rc} = n_c \sum_r \bar{x}_r = n_r \sum_c \bar{x}_c = n_r n_c \bar{x} \tag{8.21}$$

the average within column $c$ of $x_{rc} - \bar{x}_r$ is $\bar{x}_c - \bar{x}$. The average of this over all columns is zero. The variance (equation 8.20) thus splits into

$$n_r \sum_c (\bar{x}_c - \bar{x})^2 + \sum_c \sum_r (x_{rc} - \bar{x}_r - \bar{x}_c + \bar{x})^2$$

and so the total is

$$V = n_c \sum_r (\bar{x}_r - \bar{x})^2 + n_r \sum_c (\bar{x}_c - \bar{x})^2 + \sum_c \sum_r (x_{rc} - \bar{x}_r - \bar{x}_c + \bar{x})^2. \tag{8.22}$$

The partition of the variance is accomplished. Note that this is symmetrical in $r$ and $c$.

Having split the variance in this way, each component can be divided by the appropriate number of degrees of freedom: $n_r - 1$ for the rows, $n_c - 1$ for the columns, and $n_r n_c - n_r - n_c + 1$ for the residuals, for comparison. The row and column averages and the total average can be neatly added to the table.

*Example Analysing the variance*

| | Student 1 | Student 2 | Student 3 | Student 4 | Apparatus average |
|---|---|---|---|---|---|
| Apparatus 1 | 3.01 | 3.03 | 3.00 | 3.04 | 3.020 |
| Apparatus 2 | 2.93 | 2.96 | 2.95 | 3.00 | 2.960 |
| Apparatus 3 | 3.00 | 3.01 | 2.99 | 3.02 | 3.005 |
| Student average | 2.980 | 3.000 | 2.980 | 3.020 | Total 2.995 |

Some further simple arithmetic gives the variance of the student averages (columns) as 0.00110. The variance of the three apparatus averages (rows) is 0.00195. The total variance for all twelve values is 0.0119. The residual term can then most simply be evaluated by using equation 8.22:

$$V_{\text{resid}} = 0.0119 - 4 \times 0.00195 - 3 \times 0.00110 = 0.0008.$$

Thus for students, sets of apparatus, and residuals we have the estimated variances (dividing by the numbers of degrees of freedom, namely 3, 2, and 6) of 0.000367, 0.000975, and 0.000133. The ratio of students to residuals is 2.76, and Table 8.3 says that 3.29 is needed for significance even at the 10% level. For the apparatus the ratio is 7.33, exceeding the 5% critical value of 5.14. There is a difference between the sets of apparatus (at the 5% level) but the students are consistent.

This can be extended to a third factor in a very neat way, if the numbers of rows and columns are equal, $n_r = n_c = n$, and this number is also equal to the number of possible values of the third factor $t$ (often called the *treatment*). The table is thus a square—called a *Latin square* for historical reasons going back to Euler's treatment of magic squares.

If the $n$ different values of $t$ are arranged within the square so that each value of $t$ occurs once in each row and once in each column, then the third summation on the right-hand side of equation 8.22 can be further subjected to a partition of the variance into a term $n\sum_t(\bar{x}_t - \bar{x})^2$ due to differences in the treatment and a residual term, with nice cancellations like equation 8.21.

### ★ 8.5.3 Contingency Tables

Suppose you have a data sample, each member of which has two non-numeric measurements. (This can be extended to numeric ones by binning if desired.) The question asked is whether there is any correlation between the two measurements. (Notice that this is one of the rare methods available for non-numeric data.)

From the data we find the probability for each possible value of the two measurements, assuming that they are independent. These are then used to give the predicted numbers of each joint measurement, under the null hypothesis that they are independent. These predictions are then compared with the actual numbers, drawn up in a matrix (called the *contingency table*), to give a $\chi^2$ which can be tested as desired. The appropriate number of degrees of freedom is one less than the number of entries in the table, as the total number of entries is given.

*Example Failure of circuits*

Fred and Jim are engaged in the prototyping of a new and very complex item of apparatus. Fred builds eleven of them and Jim builds nine. Six of Fred's output pass their acceptance test, but five fail to perform to specification. Of Jim's output, all nine pass and none fail. Is there any evidence for a difference in performance between Fred and Jim?

We draw up the contingency table for the raw data:

| | Fred | Jim |
|---|---|---|
| Pass | 6 | 9 |
| Fail | 5 | 0 |

Taking the two outputs together, the overall probabilities for an item to pass or fail its acceptance tests are 75% and 25% respectively. Grouping together the passes and fails, an item has a probability of 55% that it was built by Fred and 45% that it was built by Jim. If these probabilities are independent the expected numbers for the table are given by simple multiplication: thus the number of failures expected from Fred is $20 \times 0.25 = 0.55 = 2.75$

Prediction under $H_0$:

| | Fred | Jim |
|---|---|---|
| Pass | 8.25 | 6.75 |
| Fail | 2.75 | 2.25 |

From the differences in the four numbers we form a $\chi^2$ (cf. equation 8.6):

$$\chi^2 = \frac{(8.25 - 6)^2}{8.25} + \frac{(9 - 6.75)^2}{6.75} + \frac{(5 - 2.75)^2}{2.75} + \frac{(0 - 2.25)^2}{2.25} = 5.45.$$

The appropriate number of degrees of freedom is one less than the number of entires, which makes three. Table 8.1 gives the critical value for 10% significance as 6.25, and our value is less. There is thus a probability of more than 10% that this difference could have arisen by pure chances, and no significant evidence for any difference between the performance of the two manufacturers.

## 8.6 PROBLEMS

*8.1*

Calculate $\chi^2$ and the number of degrees of freedom for the data on Prussian horsemen in section 3.3.1.

*8.2*

Consider the data on the number of detected neutrinos in section 3.3.1. Adopting the null hypothesis that they are described by a Poisson distribution (whose mean is a free parameter), find the $\chi^2$ and probability

(a) for all the data

(b) for all the data, excluding the interval with nine events.

Hence justify the assertion that the background is Poisson with a mean of 0.77.

*8.3*

Tests are made of a detection system which is claimed to be 'at least 90% efficient'. What can be said about this claim if

(a) It fails the first test.

(b) It fails the first three tests.

(c) It fails three of the first ten tests, and passes seven.

(d) Thirty of the first 100 tests fail.

How many tests should be made to establish the truth (or otherwise) of the claim, at some meaningful level?

*8.4*

Ten temperatures are measured, each with an error of 0.2 K:

10.2 10.4 9.8 10.5 9.9 9.8 10.3 10.1 10.3 9.9

It is suggested that they are all the same true value, differences being due to the measurement errors. Find the number of degrees of freedom and $\chi^2$. What do you conclude?

How would things be different if the original suggestion were that they are all the same true value of 10.1 K?

*8.5*

If 1000 measurements are grouped in 25 bins and fitted to a curve which is the sum of an arbitrary Gaussian on an arbitrary flat background, how many degrees of freedom are there?

★*8.6*

If a sample of data is histogrammed in a bin of width $w$, the variance of the sample is estimated from the bin contents:

$$\hat{V} = \frac{1}{N} \sum_{j=\text{all bins}} n_j (x_j - \bar{x})^2.$$

Show that this underestimates the variance of the sample, and justify the addition of *Sheppard's correction*

$$\frac{w^2}{12}.$$

*When you can measure what you are speaking about, and express it in numbers, you know something about it; but when you cannot measure it, when you cannot express it in numbers, your knowledge is of a meagre and unsatisfactory kind.*

—*Lord Kelvin*

CHAPTER 

# Ranking Methods

Unfortunately you cannot always follow Kelvin's advice; sometimes your data cannot be expressed as hard numbers.

For example, though the relative merits of composers may be hotly debated, Beethoven is certainly a greater composer than Brahms, who in turn outranks Szymanowski. But this cannot be quantified in any definite way (record sales?, number of performances?). It cannot be 'expressed in numbers' in the way that Kelvin recommends.

Similar considerations apply to painters, paintings, figure skating, and beauty contests. The items may be meaningfully *ranked* in order, but even if the ranks are expressed as numbers (Beethoven 1, Brahms 2, Szymanowski 3) these are not 'numeric' in that you cannot do arithmetic with them—what would $1 + 2 = 3$ mean in this context?

Although it is unlikely that you will be required to run a beauty contest in your professional career, it is very likely that you will be required to assess the work of colleagues, or the performance of pupils. Similar considerations apply; numbers must not be pushed too far. It is presumably true that a pupil scoring 100% in an exam performs better than one scoring 50%. However, to say that two pupils scoring 50% were equivalent to one pupil scoring 100%, in the same way as two 50 g weights are equivalent to one 100 g weight, would be obvious nonsense. Yet when people familiar only with the numerical techniques of statistics are confronted with this sort of problem, they have an urge to perform complicated and meaningless arithmetic on the pseudo-numerical data. This is dangerous as it claims misleading authority for the results.

*A precise and universally acceptable definition of the term 'nonparametric' is not presently available.*

—*John E. Walsh:* Handbook of Nonparametric Statistics

## ★9.1 NON-PARAMETRIC METHODS

Ranking methods can be useful even with genuinely numeric data, as they avoid making assumptions about the underlying distribution (and, in particular, whether it is a Gaussian or not). They do this at the price of some loss of power. Tests that do not depend, or depend only slightly, on the form of the underlying distribution are called *distribution free tests* and *robust tests* respectively. The term *non-parametric statistics* is also used, but it appears to mean different things to different people. Ranking methods might better be called *non-metric statistics.*

### ★9.1.1 The Sign Test for the Median

The median (see section 2.3.2) plays an important role in ranking methods, just as the arithmetic mean does with numerical work. It is the 'average' in the sense that half the data lies above it and half below. One simple test that can be used for it is the *sign test*, as shown in the following example. Observe how numerical tests can be invoked and validly applied, even though the data are ranked.

*Example The sign test*

Suppose the average (median) income for graduates of a certain age is £14650. Eight such graduates with degrees in accountancy earn £15100, £15500, £16950, £10320, £16330, £17100, £15160, and £16890. Is there any evidence that they are better off than average?

Seven of the eight figures are above the proposed median, and one is below. If the median is really £14650, this implies a 50:50 chance of any one of these actual numbers being above or below; the chance of a 7:1 split is (using the binomial distribution) 3.1%. Adding the one in 256 probability of them all being higher, the probability of one or less above the median is 3.5%. (A one-tailed test is used as there is no suggestion that this group might be worse off.) So there is evidence at the 5% level from this test that these accountants are more affluent than their contemporaries.

Note the good and bad features of such a test:

1. Only the sign of the difference is used so the test is independent of any assumptions about the shape of the true distribution (good feature).
2. Only the sign of the difference is used so one is throwing away data, and the test is not very powerful (bad feature).

## ★9.2 TWO RANKED SAMPLES

The question 'are these two samples the same?' is often asked of ranked data, particularly in the case where two species of item have been ranked in order together. As described in section 8.4.5, the run test can be used for such an application, as it uses only the ranking order of the data. Kolmogorov's test, as described in the same section, is also completely distribution-free. Both tests are entirely general and correspondingly weak.

### ★9.2.1 The Mann–Whitney Test

This is a distribution-free ranking test, but instead of being general like the run test it asks the more restricted question 'Are the medians of the two distributions the same?' It is therefore more powerful for this purpose. It goes under several aliases, also being known as the *U test*, the *rank sum test*, and *Wilcoxon's test*† It works as follows:

1. Suppose there are two samples, with values $x_1, x_2, x_3, \ldots, x_{N_x}$ and $y_1, y_2, y_3, \ldots, y_{N_y}$. Rank them in order together. This gives a sequence like

$$xyyxxyx.$$

2. For each $x$ value, count the number of $y$ values that come after it. Thus in the above example, the first $x$ precedes three $y$ values, the second one, the third one, and the fourth none.
3. Form the total $3+1+1+0$ and call it $U_x$, the number of times an $x$ preceds a $y$. In the same way, find $U_y$—here $3+3+1=7$. Check that

$$U_x + U_y = N_x N_y. \tag{9.1}$$

Under the null hypothesis that the averages are the same, one expects $U_x = U_y = \frac{1}{2} N_x N_y$, as each $x$ values will on average have half the $y$ sample behind it and the other half in front. If the medians are significantly different, say $x$ is ahead of $y$, then the $x$ values will precede more than their fair share of $y$ values, and $U_x$ will be greater than $U_y$. For small samples the significance is given by tables. For large samples, the Gaussian approxi-

†The Wilcoxon matched pairs test (see section 9.2.3) is different.

mation can be used, with the mean of $U_x$ equal to $\frac{1}{2}N_xN_y$, and variance $\frac{1}{12}N_xN_y(N_x+N_y)$.

An alternative method, which is completely equivalent but may be more convenient to compute, is to find the total rank for the $x$ data—here it is $R_x = 1+4+5+7 = 17$ – and then

$$U_x = N_xN_y + N_x(N_x+1)/2 - R_x. \qquad (9.2)$$

Tied ranks (an $x$ and $y$ which are equal) are a problem. Techniques for handling them are specialised. The presence of tied ranks makes the variance smaller, and is thus an error of the 'safe' side.

There is an extension of the Mann—Whitney test to the problem of several samples, called the *Kruskal–Wallis test*.

### ★9.2.2 Matched Pairs

Matching pairs (see section 8.4.3) improves the power of the two-sample problem. Again, the sign test can easily be applied.

*Example The sign test for matched pairs*
Seven children are given a vitamin supplement. The 'social behaviour index' is measured by a psychologist before and after the treatment:

| Child | A | B | C | D | E | F | G |
|---|---|---|---|---|---|---|---|
| Before treatment | 15 | 23 | 32 | 29 | 28 | 24 | 13 |
| After treatment | 22 | 26 | 38 | 33 | 30 | 28 | 10 |

Applying the sign test, improvement occurred in six out of seven cases. The null hypothesis is that the performance is equally likely to get better or worse and on this hypothesis the probability of an improvement in as many as six cases is $2^{-7}(7!/6!1! + 7!/7!0!) = 6.25\%$. So the improvement is significant at the 10% level, though not at the 5% level.

The numbers quoted in this example (15, 23, 32, etc.) appear to have high precision. Whether this is warranted is impossible to say without knowing details of the test, but one would suspect that anything that is this indefinable could not be quantified very validly. On that basis one plays safe and only looks at the sign of the change, not at its magnitude. In similar experiments measuring different quantities one might place more credence on the actual numbers. Yet even then there are hidden pitfalls. Suppose the numbers gave the accurately measured weights of the children (in some arbitrary units). Now, D and F have both gained four units. However, a weight gain of four units for F, who started with 24, is a larger *percentage* change than for D, who had 29. So are the gains of D and F to be regarded as the same, or should the gain of F be regarded as greater? It is possible that you have a definite answer to this question in a particular experiment, but if you have not, you have to fall back on using the sign test and ignoring the numerical values, even though they are hard numbers of which Kelvin would approve.

### ★9.2.3 Wilcoxon's Matched Pairs Signed Rank Test

This is more powerful than the simple sign test (section 9.2.2), and requires more validity from the numbers: it uses the magnitudes of the differences as well as the signs, but does not attach any meaning to the figures themselves. An improvement of 4 carries more weight than one of 2, though not necessarily twice as much.

1. For each pair, write down the difference. Any pairs with zero difference get thrown away. Call the remaining number $N$.
2. Rank these differences in order of magnitude (i.e. smallest has lowest rank, and ignore the signs). For equal (tied) values, give each the average of the tied ranks.
3. Add the ranks for positive and negative differences separately. Check that they add to give $\frac{1}{2}N(N+1)$. Call the smaller sum $T$.

On the null hypothesis we would expect the two sums to be 'about the same'. If the smaller sum is small enough, this refutes the null hypothesis—this can be established from Table 9.1. The entries show the value of $T$ required to achieve a given two-tailed significance level, for the appropriate number of pairs $N$. The significance is achieved if your actual $T$ is equal to *or less than* this. For a one-tailed test, the significance level is half that given in here.

TABLE 9.1
SIGNIFICANCE TABLE FOR WILCOXON'S MATCHED PAIRS

| $N$ | 0.10 | 0.05 | 0.02 | 0.01 | $N$ | 0.10 | 0.05 | 0.02 | 0.01 |
|---|---|---|---|---|---|---|---|---|---|
| 5 | 0 | — | — | — | 21 | 67 | 58 | 49 | 42 |
| 6 | 2 | 0 | — | — | 22 | 75 | 65 | 55 | 48 |
| 7 | 3 | 2 | 0 | — | 23 | 83 | 73 | 62 | 54 |
| 8 | 5 | 3 | 1 | 0 | 24 | 91 | 81 | 69 | 61 |
| 9 | 8 | 5 | 3 | 1 | 25 | 100 | 89 | 76 | 68 |
| 10 | 10 | 8 | 5 | 3 | 26 | 110 | 98 | 84 | 75 |
| 11 | 13 | 10 | 7 | 5 | 27 | 119 | 107 | 92 | 83 |
| 12 | 17 | 13 | 9 | 7 | 28 | 130 | 116 | 101 | 91 |
| 13 | 21 | 17 | 12 | 9 | 29 | 140 | 126 | 110 | 100 |
| 14 | 25 | 21 | 15 | 12 | 30 | 151 | 137 | 120 | 109 |
| 15 | 30 | 25 | 19 | 15 | 31 | 163 | 147 | 130 | 118 |
| 16 | 35 | 29 | 23 | 19 | 32 | 175 | 159 | 140 | 128 |
| 17 | 41 | 34 | 27 | 23 | 33 | 187 | 170 | 151 | 138 |
| 18 | 47 | 40 | 32 | 27 | 34 | 200 | 182 | 162 | 148 |
| 19 | 53 | 46 | 37 | 32 | 35 | 213 | 195 | 173 | 159 |
| 20 | 60 | 52 | 43 | 37 | 36 | 227 | 208 | 185 | 171 |

*Example Wilcoxon's matched pairs*
Take the set of matched pairs in the example of the previous section:

| | | | | | | | |
|---|---|---|---|---|---|---|---|
| Before treatment | 15 | 23 | 32 | 29 | 28 | 24 | 13 |
| After treatment | 22 | 26 | 38 | 33 | 30 | 28 | 10 |
| Difference | +7 | +3 | +6 | +4 | +2 | +4 | −3 |
| Rank of difference | 7 | $2\frac{1}{2}$ | 6 | $4\frac{1}{2}$ | 1 | $4\frac{1}{2}$ | $2\frac{1}{2}$ |

The sum of ranks for positive differences is $7 + 2\frac{1}{2} + 6 + 4\frac{1}{2} + 1 + 4\frac{1}{2} = 25\frac{1}{2}$. For negative differences it is $2\frac{1}{2}$. This is smaller and so $T = 2\frac{1}{2}$.

The table shows that we would need a $T$ of 3 or less for a 5% significance—so the improvement is significant at the 5% level.

## ★9.3 MEASURES OF AGREEMENT

### ★9.3.1 Spearman's Correlation Coefficient

Spearman's correlation coefficient is the equivalent for ranked data of the correlation coefficient described in section 2.6 for numerical data (which is actually called *Pearson's correlation coefficient*).

Given a data sample $\{(x_1, y_1), (x_2, y_2), \ldots, (x_N, y_N)\}$, rank the $x_i$ in order, and likewise rank the $y_i$. For each pair, find $D$, the difference between the $x$ rank and the $y$ rank. Then

$$\rho = 1 - \frac{6 \sum_{i=1}^{n} D_i^2}{N^3 - N}. \tag{9.3}$$

Like Pearson's $\rho$, it ranges between 1 for complete correlation (when all the $D_i$ are zero), through 0 for no correlation, to $-1$ for complete anticorrelation. The significance can be calculated for large $N(\geqslant 10)$ by applying Student's $t$. The $t$ variable is

$$t = \rho \sqrt{\frac{N-2}{1-\rho^2}} \tag{9.4}$$

and the number of degrees of freedom is $N - 2$.

*Example Two races*
Ten runners compete in two races. Their positions are given below—are they correlated?

(Note that the original information, the timings, is not supplied and so we have no choice but to use Spearman's $\rho$ rather than Pearson's.)

| Runner | A | B | C | D | E | F | G | H | I | J |
|---|---|---|---|---|---|---|---|---|---|---|
| Position in first race | 2 | 1 | 3 | 4 | 6 | 5 | 8 | 7 | 10 | 9 |
| Position in second race | 3 | 2 | 1 | 4 | 6 | 7 | 5 | 9 | 10 | 8 |
| Difference | 1 | 1 | −2 | 0 | 0 | 2 | −3 | 2 | 0 | −1 |
| $D^2$ | 1 | 1 | 4 | 0 | 0 | 4 | 9 | 4 | 0 | 1 |

$$\rho = 1 - \frac{6 \times 24}{1000 - 10} = 0.85.$$

This is obviously a large, positive correlation. In this case, $t = 4.55$, which for 8 degrees of freedom with a two-tailed test is safely greater than the 1% critical value of 3.355, so the correlation can be taken as significant.

### ★9.3.2 Concordance

The above example can be regarded as two tests of the runner's ability by different processes. The large $\rho$ shows these tests are consistent. For such orderings on merit ('judgements') one wants to be able to express the consistency or *concordance* between the judges. If $n$ judges place $N$ items in order of merit, one can form the coefficient of concordance, $W$, as follows:

1. Find the total rank for each item.
2. Subtract it from the average, $n(N+1)/2$, to get $D$.
3. Form

$$W = \frac{12\sum_{i=1}^{N} D_i^2}{n^2(N^3 - N)}. \tag{9.5}$$

This is constructed to go from zero (when the orderings disagree completely) to 1 (when the orderings are all the same).

Its significance can be tested with the $F$ distribution. Form

$$W' = \frac{12[(\sum_{i=1}^{N} D_i^2) - 1]}{24 + n^2(N^3 - N)}. \tag{9.6}$$

Then use the $F$ test with $F = (n-1)W'/(1 - W')$ and the numbers of degrees of freedom are $(N-1) - 2/n$ for the numerator and $(n-1)(N-1-2/n)$ for the denominator. These will not in general be whole numbers and you have to interpolate.

*Example Concord?*
Three judges arrange five items in order as follows:

| Item | A | B | C | D | E |
|---|---|---|---|---|---|
| Judge 1 ranking | 1 | 2 | 3 | 4 | 5 |
| Judge 2 ranking | 2 | 1 | 4 | 5 | 3 |
| Judge 3 ranking | 1 | 3 | 2 | 4 | 5 |

Sum the columns to get the total ranks:

| Total ranking | 4 | 6 | 9 | 13 | 13 |
|---|---|---|---|---|---|

Subtract these from the average, which is 9:

| | | | | | |
|---|---|---|---|---|---|
| $D$ | −5 | −3 | 0 | 4 | 4 |
| $D^2$ | 25 | 9 | 0 | 16 | 16 |

The total $D^2$ is 66, giving $W = (12 \times 66)/(9 \times 120) = 0.73$. So the degree of concordance between the judges is very high.

## ★9.4 PROBLEMS

*9.1*
Prove equations 9.1 and 9.2

*9.2*
The net weight of potato crisps in packets from two different manufacturers as follows:

| | | | | | | | | | |
|---|---|---|---|---|---|---|---|---|---|
| Manufacturer A | 41.2 | 40.8 | 38.8 | 42.0 | 41.6 | 38.4 | 39.8 | 41.1 | 38.1 |
| Manufacturer B | 38.9 | 38.5 | 40.5 | 37.3 | 36.9 | 40.4 | 40.6 | 37.6 | |

Apply (a) the run test and (b) the Mann–Whitney test. What can you conclude?

*9.3*
The accurately measured values of samples of resistors marked '1000 Ω' from two different manufacturers are as follows:

| | | | | | | | | | | |
|---|---|---|---|---|---|---|---|---|---|---|
| Manufacturer A | 910 | 950 | 1050 | 1060 | 940 | 1070 | 1090 | 930 | 910 | 1060 |
| Manufacturer B | 960 | 1040 | 980 | 1010 | 1000 | 1020 | 990 | 990 | 1010 | 970 |

Apply (a) the run test and (b) the Mann–Whitney test. What can you conclude?

*9.4*
The political views of fifteen husbands and wives (measured on some scale on which Karl Marx would score 0 and Attila the Hun would score 200) are as follows (each column is a husband-wife pair):

| | | | | | | | | | | | | | | | |
|---|---|---|---|---|---|---|---|---|---|---|---|---|---|---|---|
| Husband | 80 | 97 | 110 | 94 | 120 | 77 | 84 | 80 | 87 | 93 | 120 | 101 | 82 | 94 | 100 |
| Wife | 85 | 96 | 101 | 93 | 122 | 78 | 84 | 110 | 79 | 92 | 110 | 119 | 91 | 97 | 111 |

(a) Calculate Pearson's correlation coefficient.
(b) Calculate Spearman's correlation coefficient.
(c) Apply Wilcoxon's matched pairs signed rank test.

Discuss the results. In particular, explain what questions the three calculations are attempting to answer.

*The process of preparing programs for a digital computer is especially attractive because it not only can be economically and scientifically rewarding, it can also be an aesthetic experience much like composing poetry or music.*

—*Donald E. Knuth*

CHAPTER 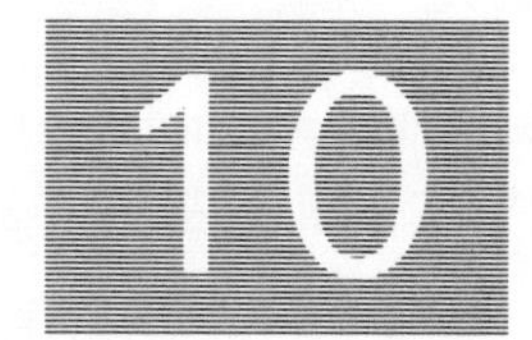

# Notes for Number Crunchers

Numerical methods are an art rather than a science, and acquired as a set of somewhat disconnected methods, clever tricks, and recipes, not as a gloriously complete subject. Some of the things I have learnt the hard way through painful experience are given here, in the hope of softening the blow when you meet the same problems.

For the sake of illustrations it will often be imagined that the work is carried out to some fixed number of significant decimal figures. A computer will of course work in binary figures (or hexadecimal), but this does not affect the principle. The points discussed are completely general to any computer or language, or indeed to calculations with a push-button calculator.

## 10.1 SIGNIFICANCE

You probably know the precision to which you want your answer. It is dictated, among other things, by the precision of the original data. However, significant figures get lost in the calculations. Even rules of thumb like 'work to five significant figures if you want two', that you sometimes see advocated, often fail to work. Retaining significance requires skill and cunning, not blind application of rules.

### 10.1.1 Subtraction

Suppose the sum

$$1000004 - 1000003$$

is performed on a machine with five decimal places of precision. It rounds the two numbers to five places, and the sum becomes

$$1.0000 \times 10^6 - 1.0000 \times 10^6$$

which gives zero.

The first commandment for retaining significance is thus: thou shalt not subtract two numbers of equal magnitude. Even though five places may sound ample (most computers work to six or seven places in standard precision) they can very easily be lost in such calculations.

Often numbers of similar magnitudes occur because some large constant has been added to both, and all you have to do is readjust the algebra. Suppose you have two points, about 10 to 20 m above sea level, and are concerned with their difference in height. Algebraically, there is nothing to stop you measuring their heights relative to the centre of the earth; indeed it may be a more physically appropriate quantity in this particular case. But adding $6.4 \times 10^6$ to each number before subtraction would be computational suicide.

Another common instance is the subtraction of two squares: if you want to evaluate $(x^2 - y^2)$ it is always better to code this as (X + Y)*(X - Y) rather than X*X - Y*Y, as the relative difference between $x^2$ and $y^2$ is smaller than that between $x$ and $y$.

*Example Significance of squares*
Consider

$$1.234^2 - 1.233^2.$$

Computing this as given leads to

$$1.522756 - 1.520289 = 0.002467$$

showing that to obtain four places of precision you have to work to seven places in the intermediate steps. The extra three places are needed to take care of the 1.52 at the front of both numbers. Writing it the other way gives

$$(1.234 + 1.233) \times 0.001 = 2.467 \times 0.001 = 0.002467$$

which never goes beyond four figures. (It can also be done in one's head, which can sometimes be used to impress colleagues!)

### 10.1.2 Computing the Standard Deviation

It was demonstrated in section 2.4.1 that the variance (standard deviation squared) of a set of data can be written in two equivalent ways:

$$V(x) = \overline{x^2} - \bar{x}^2$$
$$V(x) = \overline{(x - \bar{x})^2}.$$

In terms of actual calculations there are important differences. Evaluating the first formula just involves one pass through the data, to form the average of $x$ and of $x^2$; these are then used to give the desired result. For the second formula one has to go through the data twice, first to find the average $x$ and then again to average the squared deviations from this average. If your data are easily accessible in an array in memory then this does not matter much, but if they are written on a file (or typed in by hand) this could be a major consideration.

However, the first formula has significance problems from which the second is immune: if the magnitude of the mean is large compared to the standard deviation, the quantities $\bar{x}^2$ and $\overline{x^2}$ can be very similar, so that their subtraction should be avoided.

*Example Computing the variance*

Take the set of numbers 3, 6, 9. Then $\bar{x} = 6$ and $\overline{x^2} = (9 + 36 + 81)/3 = 42$. The variance is then given by $42 - 36 = 6$ or by $(3^2 + 0^2 + 3^2)/3 = 6$.

Now consider the set 13, 16, 19. The second method gives the same sum as before, but the first gives $262 - 256 = 6$. Three figures are needed for the final one figure precision.

For 1003, 1006, 1009 the sum is $1012042 - 1012036 = 6$. Seven figures are necessary—and if your computer does not work at this precision, or if you are calculating by hand and assume that working to, say, five significant figures is ample accuracy, then the results will be wrong.

If you are really need to avoid precision problems, and want to make only one loop over the data, this can be done by taking a working mean $x_0$ which you believe to be reasonably close to the true one. Calculate $\overline{(x - x_0)^2}$ and also $\bar{x}$ in one loop, and then use

$$\overline{(x - x_0)^2} = \overline{(x - \bar{x})^2} + (\bar{x} - x_0)^2.$$

### 10.1.3 Addition

When you add two numbers, if one is much bigger than the other then the precision of the smaller is lost:

$$1000000 + 1 = 1000000 \quad \text{to five significant figures.}$$

The second commandment for retaining significance is thus: thou shalt not add two numbers of different magnitudes.

This matters when summing or averaging a set of data. If there are many members then a loop over statements like

$$\text{SUMX} = \text{SUMX} + \text{X(I)}$$

will eventually lead to significance errors.

It is mathematically equivalent and numerically better to use the running average:

$$\text{AVGX} = \text{AVGX} + (\text{X(I)} - \text{AVGX})/\text{I}.$$

*Example Forming an average*
Suppose you work to six significant figures, and add together a very long list of numbers, all of which (for the sake of clarity) happen to be equal to 1. The sum goes $1, 2, 3, 4, \ldots$ upto 999999. The millionth item takes this to 1000000, but the next item does not affect the total, as

$$1000000 + 1 = 1000000 \quad \text{to six significant figures}$$

and neither do any items thereafter. No matter how many more items are in the list, the total will remain the same.

The running average, on the other hand, is always, quite correctly, equal to 1.

### 10.1.4 Quadratic Equations

The usual quadratic formula

$$x = \frac{-b \pm \sqrt{b^2 - 4ac}}{2a}$$

gives significance errors for one of the roots if $b$ is much greater than $4ac$. This can be avoided writing

$$q = \tfrac{1}{2}[b + \text{sgn}(b)\sqrt{b^2 - 4ac}]$$

and the roots are then $q/a$ and $c/q$.

## 10.2 OTHER DO'S AND DON'T'S

(Mostly don't's)

### 10.2.1 Matrix Inversion

The general advice to people who want to invert a matrix is *don't*. Sets of linear equations can be solved without explicitly inverting the matrix: this is faster by a factor of about three (and also more accurate). If you do have to invert a matrix, don't do it by evaluating the cofactors—there are many better techniques. Consults a good library of routines (e.g. that of the Numerical Algorithms Group) rather than writing your own.

### 10.2.2 Fitting Curves

Your first guess at fitting a curve would probably be $a + bx + cx^2 + dx^3 \ldots$ adjusting $a, b, c, d$. This is not usually the best choice. Use a proprietary polynomial instead.

Chebyshev polynomials have some very nice properties of orthogonality and convergence, and are probably the best polynomials around. $x$ should be in the range between $-1$ and $+1$. Then fit

$$a + bT_1(x) + cT_2(x) + dT_3(x)$$

where

$$T_1 = x$$

$$T_2 = 2x^2 - 1$$

$$T_3 = 4x^3 - 3x.$$

For higher orders, use $T_{n+1}(x) = 2xT_n(x) - T_{n-1}(x)$. Note that definitions given in the literature of $T_0$ differ.

It may well be stupid to try and fit a single function over the whole range, and more appropriate to split the range into intervals and fit different functions in each. These different functions then have to join up smoothly at the interval edges—such functions are called *splines*. Again, this is a well-established technique, and a look at the specialist literature will be worth while.

## 10.3. RANDOM NUMBER GENERATION

Most computers have a generator that produces random numbers between 0 and 1, in a uniform distribution, called something like RAN(DOM). This can trivially be adjusted to any other range: e.g. to generate a random direction in three dimensions

PHI = 6.28315*RAN(DOM) THETA = ACOS(2.0*RAN(DOM) − 1.0).

If $y$ is a function of $x$ and $x$ has a uniform distribution, then the probability distribution for $y$ is proportional to $|\mathrm{d}x/\mathrm{d}y|$. For example, if $y = -k \ln(x)$ then $\mathrm{d}y/\mathrm{d}x = -k/x$, and $P(y)$ is proportional to $x$, which is $\mathrm{e}^{-y/k}$. So this generates an exponential distribution.

T = −TAU*LN(RAN(DOM)) generates values of $t$ according to an exponential distribution of mean $\tau$.

The number of distribution functions that can be generated in this way is very limited. The alternative is weighted rejection. For example, to generate a set of numbers in the range 0–1 according to $P(x) = 2x$:

1. Generate a uniformly random $x$. Find $P(x)$.

2. Divide this by the maximum possible weight—here 2.0. If you do not know this, overestimation is safe.
3. Compare this ratio with another uniformly generated random number. If the ratio is larger than the random number, accept it. Otherwise go back to step 1 and try again.

This is then repeated many times. Many values are generated only to be thrown away, so this can be inefficient. The efficiency can be increased by generating an initial distribution known to be close to that desired, and loading the weight approximately.

Such random numbers are used in *Monte Carlo* techniques. For example, suppose you are measuring photons, emitted with energies which are random, but with a known distribution. You use a certain apparatus, the efficiency of which depends on the photon energy in a known way. You can find the overall efficiency by generating a large number of photons with energies selected according to the known random distribution, and applying the known efficiencies.

If your computer time is limited, the accuracy with which you know the efficiency may be limited by the number $N$ of 'Monte Carlo' photons you can generate. In the language of estimation (Chapter 5) you have taken $N$ samples of the mean efficiency, and thus have a known accuracy inversely proportional to $\sqrt{N}$. The accuracy may be improved (cf. section 5.7) if the events are not generated completely at random, but with an even spread.

*Code—a set of symbols whose primary purpose is to restrict comprehension.*

—*Webster's 3rd International Dictionary*

## 10.4 STYLE

Good programming style is important, not just a luxury. Programs are organic; a good program lives a long time, and has to be changed to meet new uses; a well-written program makes such adaptations easier. A logical, clearly written program helps greatly when tracking down possible bugs. However, although there is a lot of good advice on style laid down by various people, there are no hard and fast rules, only guidelines. It is like writing English in this respect.

Comment statements are essential, but you can have too many. An overindulgence in flashy banners and uninformative comments can obscure the actual code; a simple program or routine can be spread over several sheets of paper instead of fitting neatly on one page where you can see at a glance what it does.

A line of code should generally be self-explanatory. If it is not, because you are being clever, or for any other reason, then it needs a comment. A good programmer should be able to write clear statements that require no extra description. (That's my story, anyway!)

On the other hand, *sections* of code are not usually self-explanatory, and should be commented at the start. This also breaks up the code into logical units, so you can see where one activity stops and the next starts. It is like separating English prose into paragraphs.

In all but exceptional cases, do not worry about the speed and efficiency of your program. It is not worth a week of your time shaving 3 seconds off the running time of a program. There are various tricks to use, but the compiler (if it is any good) is better than you at them.

The GOTO statement is the subject of much heated debate. It is certainly true that GOTOs can be used in such a way as to make a program very hard to read—but so can the alternative IF-THEN-ELSE construction if nested more than a couple of layers deep. GOTO-free programming is not an automatic passport to good practice. The clarity of the program depends far more one the use made of the statements by the programmer than the statements themselves.

The best way to learn to write good programs is to read other people's. Notice whether they are clear or obscure, and try and understand why. Then imitate the good points and avoid the bad in your own programs.

# Bibliography

The following list is by no means exhaustive, but represents the sources which the author has found most useful. Though given under the chapter to which they particularly apply, most of them are of much wider application.

**Chapter 1**

M. J. Moroney: *Facts from Figures* (Penguin)

A gentle introduction to probability and statistics, particularly good on the uses and limits of statistical methods. Written in the 1950s it is still going strong, despite some quaintly dated sexist language.

**Chapter 2**

Darrell Huff: *How to Lie with Statistics* (Penguin)

This delightful work is essential reading for anyone trying to survive the twentieth century.

**Chapter 3**

W. T. Eadie *et al.*: *Statistical Methods in Experimental Physics* (North-Holland, 1971)

This is a fairly advanced and specialised book, but Chapter 4 contains comprehensive details of a large number of probability distributions.

**Chapters 4, 5, 6, and 8**

M. G. Kendall and A. Stuart: *The Advanced Theory of Statistics* (Charles Griffin & Co., 3 volumes)

This is, and will probably always be, the definitive textbook on the theory of estimation, and indeed to most of the rest of statistics. When referring to it, note not only the edition, but also the difference between the early 2-volume and later 3-volume version, as the edition number become reset to one when the change occurred. References in this book are to the 3rd 3-volume edition.

A. G. Frodesen *et al.*: *Probability and Statistics in Particle Physics* (Universitetsforlaget, Bergen-Olso-Tromso, 1979)

An excellent general work; not as specialised as the title implies.

Louis Lyons: *Statistics for Nuclear and Particle Physicists* (Cambridge University Press, 1986)

Contains many useful insights and explanations.

J. Orear: *Notes on Statistics for Physicists* (University of California report UCRL-8417 (1958)-unpublished)

This invaluable set of notes has been circulating in *samizdat* form for many years. If you are fortunate enough to catch sight of a copy, grab it and run to the nearest Xerox machine.

**Chapter 7**

A. N. Kolmogorov: *Foundations of the Theory of Probability* (Chelsea Publishing Company, 1950)

This is Kolmogorov's original work. The material is covered by most mathematical textbooks on the subject of probability.

Richard von Mises: *Probability, Statistics, and Truth* (Dover Publications, 1957)

A very readable explanation of the frequency theory of probability.

David Miller: *A Pocket Popper* (Fontana, 1983)

Pages 199 to 206 contain an account of the propensity theory.

Confidence intervals are also fully discussed in the works by Eadie *et al.* and Kendall and Stuart cited above.

**Chapter 9**

A. Haber and R. P. Runyon: *General Statistics* (Addison-Wesley, 1979)

Aimed at readers in the social sciences, and correspondingly nonmathematical, but gives a good brief description of non-parametric tests.

S. Siegel: *Nonparametric Statistics for the Behavioural Sciences* (McGraw-Hill, 1956)

Contains tables of significance for the Mann–Whitney test, and many others.

**Chapter 10**

P. K. MacKeown and D. J. Newman: *Computational Techniques in Physics* (Adam Hilger, 1987)

A useful general account. Anyone wanting to do serious scientific computing should get it.

W. H. Press *et al.*: *Numerical Recipes—The Art of Scientific Computing* (Cambridge University Press, 1986)

A useful thick reference work. Anyone wanting to do serious scientific computing should make sure their library has a copy.

APPENDIX 

# Answers to Problems

*2.1* The mean is 19.26 and $\sigma$ is 0.49.
*2.2* The mean increases to 19.95 and $\sigma$ to 3.44.
*2.3* The skew is 0.23 for the students, increasing to 4.65 if the lecturer is included.
*2.4* The averages are 37.50 and 55.25, with standard deviation 25.9 and 14.2. The covariance is 207.5 and the correlation is 0.57.
*2.5* Multiply out the brackets and take averages. Some of the $\bar{x}$ terms then cancel.
*2.6* I found that bins 10 units wide gave acceptable results (specifically bins of 0–9, 10–19, 20–29,...,90–99).
*2.7* From the raw data, the mean is 61.59, the median is 63.5, and the mode is 79.
Binning as above, the mean obtained from the bins is 61.75, and the median is the 60–69 bin. The mode is also the 60–69 bin. Precision has been (needlessly) lost for the mean and median, but the mode is more sensible than the value from the raw data.
*2.8* $\sigma$ is 16.7 for the raw data and 16.8 for the binned data. The maximum (in the 60–69 bin) is 18, and the width at a height of 9 is 4 bins, each of 10 units, so the FWHM is 40. $2.35\sigma$ is 39, so the agreement with equation 2.12 is excellent—indeed, it is better than one has any right to expect.
*2.9* Consider $V(x-y)$. This is given by the expression

$$\sigma_x^2 + \sigma_y^2 - 2\rho\sigma_x\sigma_y$$

This must be non-negative, whatever the values of $\sigma_x$ and $\sigma_y$. So, writing their ratio as $u$, the quantity $u^2 - 2\rho u + 1$ must be non-negative, and if it is solved for $u$ using the quadratic formula the discriminant must be negative or zero: this gives $\rho^2 \leqslant 1$.

*3.1* 61%. 139 missiles.
*3.2* 535 missiles.
*3.3* 10.9%.
*3.4* $\lambda$ is 7.85, so $\sigma$ is 2.80, and the limit of 4.5 is 1.20 $\sigma$ from the mean. Table 3.3 gives the probability as 11.5%.

*3.5* For 60 cars, each with a probability of 99% for not giving a lift, the binomial formula gives a probability of $0.99^{60}$, which is 54.7%.

The mean number of lift-giving cars in one hour is 0.6. The probability that none will appear in a given hour is given by the Poisson formula as $e^{-0.6}$, which is 54.9%.

Please get the important difference between the two situations clear, and understand why the two different formulae are applied. Once you understand this, you are on the way to becoming a statistics expert.

*3.6* (a) 21.87%. (b) 0.75%. (c) 13.79%. (d) 67.36%. (e) 48.35%. (f) 86.70%. (g) $0.675\sigma$. (h) $2.33\sigma$.

*3.7* As the Gaussian is symmetric, all odd central moments must vanish, so the skew is zero. Table 3.1 gives an equation for higher even powers which (with $n=2$) reads

$$\int_{-\infty}^{\infty} x^4 e^{-ax^2}\,dx = \frac{1\times 3}{4a^2}\sqrt{\frac{\pi}{a}}\;.$$

showing that the fourth central moment is $3\sigma^4$; hence the curtosis is zero.

*4.1* 1 measurement at 0.2 mm. Even if the 10 are independent their combined error of $1/\sqrt{10}$ is larger than 0.2. (But 10 measurements would give a consistency check.)
*4.2* 299 793 100 ± 2900 m/sec.

Incidentally, since 1983 the speed of light has been *defined* to be 299 792 458 m/sec. So what are such experiments actually measuring?

*4.3* The third measurement, despite its small quoted accuracy, is out of line with the other four. Lacking further information, assume that it is wrong and the others are all right. Their combined value is 299 791 100 ± 1800 m/sec.
*4.4* 1570 ± 40 volts.
*4.5* 50 ± 1 mA.
*4.6* For $\theta = 0.56$ they are 0.008 for $\sin\theta$, 0.005 for $\cos\theta$, and 0.014 for $\tan\theta$.

For $\theta = 1.56$ (which is almost $\pi/2$) they are 0.0001 for $\sin\theta$, 0.010 for $\cos\theta$, and for $\tan\theta$ the change over the range of $\theta$ is so large that the 'small errors' assumption breaks down, and so sensible error can be quoted.

*4.7* Differentiating equation 4.6 gives

$$\frac{\partial \bar{x}}{\partial x_i} = \frac{1/\sigma_i^2}{\Sigma_j (1/\sigma_j^2)}.$$

Each $x_i$ has error $\sigma_i$: squaring and adding as in section 4.3.2 gives equation 4.7.
*4.8* The best result is the unweighted arithmetic mean of the two, 406 counts/minute. The first measurement is *not* more accurate than the second; the error is given by the square root of the *expected* number of counts, which is the same for both.
*4.9* If the errors come entirely from Poisson statistics then they should be combined, weighted according to their expected numbers of decays, which are proportional to their efficiencies and running times. If they come entirely from calibration uncertainties then a weighted combination should be used, with the errors given. Lacking such information, the question is unanswerable.

*4.10* The matrices **V** and **G** are

$$\begin{pmatrix} \sigma_V^2 + S_V^2 & S_V^2 & 0 \\ S_V^2 & \sigma_V^2 + S_V^2 & 0 \\ 0 & 0 & S_R^2 \end{pmatrix} \begin{pmatrix} 1/R & 0 & -I_1/R \\ 0 & 1/R & -I_2/R \end{pmatrix}$$

and the error matrix is

$$\frac{1}{R^2}\begin{pmatrix} (\sigma_V^2 + S_V^2) + S_R^2 I_1^2 & S_V^2 + S_R^2 I_1 I_2 \\ S_V^2 + S_R^2 I_1 I_2 & (\sigma_V^2 + S_V^2) + S_R^2 I_2^2 \end{pmatrix}.$$

*5.1* $N_S + N_F$ is a fixed number, $n$. From the properties of the binomial (equation 3.8) the expected number of successes is $np = (N_S + N_F)p$ This shows that the estimator is unbiased. The variance on this is $np(1-p)$ (equation 3.9), so the variance on our $\hat{p}$ is $p(1-p)/n$, which vanishes as $n \to \infty$, so the estimator is consistent.

*5.2* (a) The mean is 20.21, with error $0.8/\sqrt{7} = 0.30$. (b) The root mean squared deviation from 20 is 0.87. Using equation 5.23, the error is $0.87/\sqrt{14} = 0.23$. (c) The estimated standard deviation for the data (including Bessel's correction—see section 5.2.2) is 0.91, with the error given by equation 5.24 as $0.91/\sqrt{12} = 0.26$. (d) Using the answer to (3), our best value for the error on the mean is $0.91/\sqrt{7} = 0.34$. If you are worried that there is an 'error on the error', this is dealt with using Student's $t$—see section 7.3.

*5.3* Without loss of generality, suppose that $\mu = 0$. Expanding $\langle\overline{(x-\bar{x})^3}\rangle$ gives $\langle\overline{x^3}\rangle - 3\langle\overline{x^2}\bar{x}\rangle + 2\langle\bar{x}^3\rangle$. The first term is

$$\left\langle \frac{1}{N}\sum_i x_i^3 \right\rangle = \frac{1}{N}\sum_i \langle x_i^3 \rangle = \frac{1}{N}\sum_i \langle x^3 \rangle = \langle x^3 \rangle.$$

The expectation value in the second term is

$$\left\langle \frac{1}{N^2}\sum_i x_i \sum_j x_j^2 \right\rangle = \frac{1}{N^2}\sum_i\sum_j \langle x_i x_j^2 \rangle.$$

The $x$ values are independent, so for any $j \neq i$ the term is $\langle x_i \rangle \langle x_j^2 \rangle$, and (remembering that $\mu = 0$) $\langle x_i \rangle = 0$. Only the terms with $j = i$ remain, giving $(1/N^2)\sum_i \langle x^3 \rangle = \langle x^3 \rangle / N$.

In the same way the third term contains $\langle x^3 \rangle / N^2$, giving

$$\langle x^3 \rangle - \frac{3\langle x^3 \rangle}{N} + \frac{2\langle x^3 \rangle}{N^2} = \frac{(N-2)(N-1)}{N^2}\langle x^3 \rangle.$$

Hence the result.

*5.4* In terms of expectation values, equation 2.15 gives $\langle (x-\mu)^4 \rangle = \sigma^4(3+c)$. $\langle (x-\mu)^2 \rangle^2$ is $\sigma^4$. Putting these in equation 5.17 gives the answer. Note that equation 5.18 follows immediately.

*5.5* Suppose the measurements are $s$ and $c$. The likelihood function is

$$\exp -(s - \sin\theta)^2/2\sigma^2 \exp -(c - \cos\theta)^2/2\sigma^2.$$

Setting the differential to zero gives $s\cos\hat{\theta} = c\sin\hat{\theta}$, i.e. $\hat{\theta} = \tan^{-1}(s/c)$.

*5.6* This is a continuation of the example following equation 5.26. The second derivative of the log likelihood is $\sum(-(2t/\tau^3) + (1/\tau^2))$. Taking the expectation value gives $N/\tau^2$, as $\langle t \rangle = \tau$. Negating and inverting gives the MVB by equation 5.28, and for large $N$ this is the ML variance.

*6.1* Using equations 6.4 and 6.5, with $x \equiv t$, $y \equiv d$, $m \equiv v$, gives the answer $v = 10.1 \pm 0.2$ mm/sec. The $\chi^2$ is 3.0 which, for 5 degrees of freedom, is all right.
*6.2* This is like the previous problem except that now $d = x$, $t \equiv y$, $m \equiv (1/v)$. Using the law of combination of errors to get from the error on $1/v$ to the error on $v$, equations 6.4 and 6.5 give $v = 10.1 \pm 0.1$. The $\chi^2$ is now 10.6.
*6.3* Rewriting the formula as $t = \sqrt{2/g}\sqrt{d}$ one can identify $y$ with $t$ and $x$ with $\sqrt{d}$. Using 6.4 and 6.5 gives $g = 11.32 \pm 0.16$, and $\chi^2$ is 23.8 (for 4 degrees of freedom). Adding an arbitrary constant to the time and using equations 6.6 and 6.10 gives $g = 9.87 \pm 0.32$ and a $\chi^2$ of 3.9 for 3 degrees of freedom. The $\chi^2$ of the former is too large to be believed, whereas the latter is acceptable, indicating that the time for the field to die away is important and has to be considered. Notice that the latter method has a larger error because of the uncertainty in this time, but this does not mean the former is 'better'!

The constant comes out at 0.05 seconds, with an error showing that its value is incompatible with zero, which is another way of saying the same thing.

Plotting the points and the two curves on a graph will make the situation clear.
*6.4*

$$a = \frac{\overline{xy}\,\overline{\sin^2 x} - \overline{y\sin x}\,\overline{x\sin x}}{\overline{x^2}\,\overline{\sin^2 x} - \overline{x\sin x}^{\,2}} \qquad b = \frac{\overline{x^2}\,\overline{y\sin x} - \overline{xy}\,\overline{x\sin x}}{\overline{x^2}\,\overline{\sin^2 x} - \overline{x\sin x}^{\,2}}$$

giving $a = 1.963$, $b = 1.035$.
*6.5* Fit a straight line by plotting the time against the log of the count rate. The error on the log is, by Poisson statistics, $1/\sqrt{N}$, and the points are weighted accordingly. This gives a half-life of $1.03 \pm 0.02$ hours.

*7.1* Do not try to convince him that HT and TH are different using logic—he will run rings around you. (For particles obeying Bose—Einstein statistics he would be right!) Invite him to toss pair of coins, you pay him 2 francs if they give the same result, he pays you 3 francs if they are different, and see how long his convictions last.
*7.2* Use equations 7.1 and 7.2. The original probability, $P(\mathrm{MSF})$, is 0.01%. Then

$$P(\mathrm{MSF}\,|\,\mathrm{spots}) = \frac{1}{0.03 \times 0.9999 + 1 \times 0.0001} \times 0.0001 = 0.0033$$

$$P(\mathrm{MSF}\,|\,\mathrm{lethargy\ and\ spots}) = \frac{1}{0.1 \times 0.9967 + 1 \times 0.0033} \times 0.0033 = 0.032$$

and so on for the other symptoms. Applying them in any other order gives the same result.
*7.3* See F. James and M. Roos, *Nuclear Physics*, **B172**, p. 475. (1980)
*7.4*

$$\sum_0^3 P(r; 0.19, 8) = 0.185 + 0.348 + 0.285 + 0.134 = 0.95$$

i.e. if $p$ is 19% (or lower), the probability of 4 (or more) successes is 5% (or less). The arithmetic for the upper limit is the same.

Using the Gaussian approximation: the estimate for $p$ is 0.5, so the variance on $r$ is $0.5 \times 0.5 \times 8 = 2$. The error on the estimate is $\sqrt{2}/8 = 0.177$, and 90% limits are given by $1.645\sigma$, i.e. 21 and 79%.
*7.5* From Table 7.1, the 90% confidence interval for the number of decays is from

3.29 to 13.15. There are $6.022 \times 10^{32}$ protons in the sample, so the probability of decay in 1 year is in the range $5.46 \times 10^{-33}$ to $2.18 \times 10^{-32}$. These give the limits on the mean lifetime (and the lower number limit gives the upper lifetime limit!) as $4.6 \times 10^{31}$ to $1.83 \times 10^{32}$ years, and hence (multiplying by $\log_e 2$) half-life limits of $3.2 \times 10^{31}$ to $1.27 \times 10^{32}$ years.

*7.6* It is still true that the limits on the number of observations are 3.29 to 13.15. As $N_{\text{obs}} = N_{\text{decays}} + N_{\text{back}}$, and as the convolution of two Poissons gives another Poisson, the limits on $N_{\text{decays}}$ are 0.29 to 10.15. The half-life limits are $4.11 \times 10^{31}$ to $1.44 \times 10^{33}$ years.

*7.7* There is now no meaningful lower limit on the number of decays, and thus no upper limit on the lifetime. However, there is an upper limit of 5.15 decays, which shows that the half-life must be at least $8.1 \times 10^{31}$ years, with 95% confidence.

*7.8* The mean is 7.5 and $s$ is 1.3, so the mean has error 0.65 and differs from the usual value by 2.9$s$. Consulting Table 7.2, for 3 d.o.f. and a one-tailed test, the confidence level for this is somewhere between 95 and 97.5%. The probability that further investigations will prove disappointing is less than 5%.

*8.1* Using the mean given, equation 8.6 gives a $\chi^2$ of 0.65 for 4 degrees of freedom. This is extremely small—the probability of a value this 'good' is only 4.3%. The explanation is presumably that, of the various examples that could be found of Poisson statistics, the one with the most convincing agreement found its way into all the textbooks.

*8.2* Using all the data gives a mean of 0.777, which gives a set of predictions whose agreement with the data is described by a $\chi^2$ of 3323.6, for 9 degrees of freedom. This is quite unacceptable, and shows that the predictions and the data disagree.

Omitting the final bin reduces the mean slightly, to 0.774, and the $\chi^2$ falls to 3.2, for 8 degrees of freedom, which is quite acceptable.

*8.3* The null hypothesis is that the probability of failure is 10% or less. Thus the probability of a failure on the first test is 10% (or less), and such a failure would cause us to reject the hypothesis if we worked at the 10% level but not at any more significant level.

The probability of three failures is 0.1%, and the claim is thus rejected if you work to any reasonable significance level.

Using the binomial formula, the probability of three or more failures from ten trials (with $p = 0.9$) is 7%. Thus the claim is rejected if we work at the 10% level, but accepted if we work at the 5% level.

For 100 trials, use the Gaussian approximation. We expect ten failures (or less) with a standard deviation of $\sqrt{100 \times 0.1 \times 0.9} = 3$ (or less) if the claim is true. The result is $6\frac{2}{3}\sigma$ away from zero, the significance of which is overwhelming.

We are investigating the hypothesis of 90% efficiency. More assumptions about the tests are needed: suppose that we would not want to use a system that was only 85% efficient, whereas if it were 86% efficient we would not be worried. As the number of trials needed, $N$, will be large, use the Gaussian approximation. Suppose we choose a significance of 5% and a power of 99% for the test. If there are $r$ failures, the rejection region consists of those values of $r$ greater than $r_{\text{crit}} = 0.1 \times N + 1.645 \times \sqrt{0.09N}$. This is also the point $0.15 \times N - 2.33 \times \sqrt{0.1275N}$. This gives $N = 703$.

*8.4* The average value is 10.12 K, and $\chi^2$ is 14.9 for 9 degrees of freedom. So (see Table 8.1) the hypothesis would (just) be rejected, at the 10% level.

If the true value is 10.1 K, then $\chi^2$ is 15.0 for 10 degrees of freedom, which is (just) acceptable at the 10% level.

*8.5* $25 - 3 - 1 = 21$ degrees of freedom. The figure of 1000 is a red herring.

*8.6* Referring to equation 8.18, the desired total variance is made up of the known between-group variance and an unknown residual variance. However, if it can be assumed that the data within a bin are spread uniformly within that bin, then equation 3.25 gives its variance, and the result follows.

*9.1* Each of the total $N_x N_y$ $x-y$ pairs contributes 1 to either $U_x$ (if the $x$ comes first) or $U_y$. Hence equation 9.1.

The rank of an item is one more than the number of other items, of either species, that precede it. Thus $R_x - N_x$ gives the total number of times a $y$ precedes an $x$, plus the number of times an $x$ precedes another $x$. The former is defined as $U_y$, the latter is $(N_x-1)+(N_x-2)+(N_x-3)+\cdots+1$ which is $N_x(N_x-1)/2$. Combining these with equation 9.1 gives equation 9.2.

*9.2* The sequence is BBBAABABABBBAAAAA which has eight runs. The expected number is 9.5, by equation 8.9. So there is no significant difference there.

The total rank for A is 100, and for B is 53; hence $U_A$ is 17 and $U_B$ is 55. The expected mean is 36; the standard deviation (under the Gaussian approximation) is 10; the difference is about $2\sigma$, and therefore looks significant. Consulting a table (or use of a NAG subroutine) gives the actual significance level as 3.4%.

So, if we assume that the distributions have the same shape, the packets of manufacturer A are heavier (at the 5% confidence level).

*9.3* $U_A$ and $U_B$ are both 50, so there is no significant result from the Mann–Whitney test; however, there are only three runs, instead of the expected $11.0 \pm 2.2$ (using the large $N$ approximation, which is on the limit of its validity here, but the result is so clear-cut it does not matter). This is $3.6\sigma$, and thus there is a significant difference between the samples, though their averages are compatible.

Distributions like the values from A—the 'volcano distribution'—do occur in such situations. The original distribution will have been a normal Gaussian (roughly speaking) of mean 1000, but the manufacturer has removed the components closest to the nominal value to sell them at a higher price.

*9.4* Pearson's $\rho$ is 0.71. Spearman's $\rho$ is 0.75. The matched pairs test gives a vale of $T$ of 36, which (for $N = 14$) is not significant at any serious level.

The two correlation coefficients address the question 'Do husbands and wives tend to have the same political views', to which these data give a conclusive positive response. The two calculations make different assumptions about the meaningfulness of the numerical values given, but the conclusion is the same.

Wilcoxon's test asks 'Do (married) men and women have the same views, or does one sex tend to lie to the right (or left) of the other?', and the answer is 'There is no evidence that this is so.' By considering differences within married pairs, the large variations in sampling that would arise from variations in obviously important factors (like income, social standing, place of residence, etc.) are ironed out.

APPENDIX 2

# Proof of the Central Limit Theorem

The mathematical tools introduced here are described in some detail, as they have many other useful applications. Even so, what follows is a strictly non-rigorous exercise: for a more thorough analysis, which (for example) lays down conditions on the existence of various moments and functions, consult the specialist textbooks.

Begin by considering

$$\tilde{P}(k) = \int e^{ikx} P(x)\, dx$$

which is the Fourier transform of the probability density $P(x)$, and also the expectation value $\langle e^{ikx} \rangle$. As would be expected from its dual significance, this function is extremely useful and important in mathematical statistics, and is given the special title of the *characteristic function.* For discrete variables, one can define

$$\tilde{P}(k) = \sum e^{ikr} P(r)$$

in exactly the same way.

As an example of how useful these functions can be, notice that if it is differentiated $q$ times and the value taken at $k = 0$, this gives

$$\left.\frac{\mathrm{d}^q \tilde{P}}{\mathrm{d}k^q}\right|_{k=0} = \int (ix)^q \, \mathrm{e}^0 \, P(x) \, \mathrm{d}x = i^q \langle x^q \rangle$$

so the moments of the distribution $P(x)$ can easily be found.

Another use for characteristic functions comes in dealing with convolutions. These often occur in statistical problems, where one variable is the sum of another two (or more) random variables. If $x = x_1 + x_2$, say, and $x_1$ and $x_2$ have probability density functions $f(x)$ and $g(x)$, then the probability density for $x$, $F(x)$, has to consider all possible pairs of values that add to give $x$: it is given by the *convolution* of $f$ and $g$, sometimes written $f * g$, i.e.

$$F(x) = \int f(x') g(x - x') \, \mathrm{d}x'.$$

Characteristic functions are useful for these because *the characteristic function of a convolution is the product of the individual characteristic functions.*

*Proof:*

$$\begin{aligned}
\tilde{F}(k) &= \int \mathrm{e}^{ikx} F(x) \, \mathrm{d}x \\
&= \iint \mathrm{e}^{ikx} f(x') g(x - x') \, \mathrm{d}x \, \mathrm{d}x' \\
&= \iint \mathrm{e}^{ikx'} f(x') \mathrm{e}^{ik(x - x')} g(x - x') \, \mathrm{d}x \, \mathrm{d}x'
\end{aligned}$$

and, making the change from $\mathrm{d}x$ to $\mathrm{d}(x - x')$:

$$\tilde{F}(k) = \tilde{f}(k) \, \tilde{g}(k).$$

Reverting now to the other meaning of the characteristic function, note that the expectation value can be formally expanded as a power series in $x$:

$$\langle \mathrm{e}^{ikx} \rangle = 1 + ik \langle x \rangle + \frac{(ik)^2}{2!} \langle x^2 \rangle \cdots.$$

If you take the logarithm of this series, this can also be expanded, using $\ln(1 + \xi) = \xi - \xi^2/2 + \xi^3/3 \cdots$. As before, this is a formal expansion and all terms are kept. The result of this double expansion is a power series in $(ik)$, where the coefficient of $(ik)^r$ is of order $x^r$, as each power of $k$ goes with an equal power of $x$. Thus

$$\ln \tilde{F}(k) = (ik)\kappa_1 + \frac{(ik)^2}{2!}\kappa_2 + \frac{(ik)^3}{3!}\kappa_3 + \cdots.$$

Each $\kappa_r$ is made from expectation values of $x$, of order $x^r$. Specifically, the first few are (as you work out by doing the expansion yourself)

$$\kappa_1 = \langle x \rangle$$

$$\kappa_2 = \langle x^2 \rangle - \langle x \rangle^2$$

$$\kappa_3 = \langle x^3 \rangle - 3\langle x \rangle \langle x^2 \rangle + 2\langle x \rangle^3$$

$$\ldots$$

The $\kappa_r$ rejoice in the name of 'the semi-invariant cumulants of Thiele'. They have the nice property that, under a change of origin, $x \to x + a$, then $\kappa_1 \to \kappa_1 + a$ but all the other $\kappa_r$ are unaffected. Under a change in scale, $x \to bx$, all that happens is that $\kappa_r \to b^r \kappa_r$.

Furthermore, as the series is the logarithm of the characteristic function, and as the characteristic functions of a convolution multiply, each cumulant for the convolution of two (or more) functions is merely given by the sum of the equivalent cumulants of the functions concerned.

We now have all the tools to hand and can get on with the proof. First consider the Gaussian distribution function. Its characteristic function (i.e. Fourier transform) is also a Gaussian. This means that its cumulants are all zero apart from the first two. This shows us our target: to show that a function tends to a Gaussian it suffices to show that the larger cumulants ($r \geqslant 3$) vanish.

Now consider what happens as more and more variables are added together (cf. Figure 4.1). The distribution of the sum will be described by a distribution for which $\kappa_r$ is the sum of all the individual $\kappa_r$. Suppose that the average (over the $N$ distributions) is $\bar{\kappa}_r$; then the cumulant for the final distribution is $N\bar{\kappa}_r$. For $r = 2$, given (see above) that $\kappa_2$ is just the variance, this proves the second part of the CLT.

Now we need to show that the higher cumulants are small. Small compared to what? We have just seen that a change of scale affects them, though it cannot affect the 'Gaussian-like'—or otherwise—nature of the distribution. We can get rid of this ambiguity by rewriting the distribution as a function of $x/\sigma$, which reduces it to what should (if it is Gaussian) be unit normal form. To do this we scale $x$ by the standard deviation: for our total distribution, this is $\sqrt{N\bar{\kappa}_2}$, and we accordingly divide by this scale factor. The cumulants are affected as described above, giving, for the (scaled) total distribution

$$\kappa_r = \frac{N\bar{\kappa}_r}{(N\bar{\kappa}_2)^{r/2}}.$$

For $r = 2$ this is 1, as prescribed. For higher values of $r$ it falls with $N$ like $N^{1-r/2}$, and therefore goes to zero in the limit as $N \to \infty$. The theorem is proved.

# Index